Multi-Resolution Methods for Modeling and Control of Dynamical Systems

CHAPMAN & HALL/CRC APPLIED MATHEMATICS
AND NONLINEAR SCIENCE SERIES
Series Editors   *Goong Chen and Thomas J. Bridges*

Published Titles

Forthcoming Titles

CHAPMAN & HALL/CRC APPLIED MATHEMATICS
AND NONLINEAR SCIENCE SERIES

Multi-Resolution Methods for Modeling and Control of Dynamical Systems

Puneet Singla
John L. Junkins

 CRC Press
Taylor & Francis Group
Boca Raton London New York

CRC Press is an imprint of the
Taylor & Francis Group, an **informa** business

A CHAPMAN & HALL BOOK

Chapman & Hall/CRC
Taylor & Francis Group
6000 Broken Sound Parkway NW, Suite 300
Boca Raton, FL 33487-2742

© 2009 by Taylor & Francis Group, LLC
Chapman & Hall/CRC is an imprint of Taylor & Francis Group, an Informa business

No claim to original U.S. Government works
Printed in the United States of America on acid-free paper
10 9 8 7 6 5 4 3 2 1

International Standard Book Number-13: 978-1-58488-769-0 (Hardcover)

Library of Congress Cataloging-in-Publication Data

Singla, Puneet.
    Multi-resolution methods for modeling and control of dynamical systems / Puneet Singla and John L. Junkins.
        p. cm. -- (Chhapman & hall/CRC applied mathematics and nonlinear science ; 16)
    Includes bibliographical references and index.
    ISBN 978-1-58488-769-0 (alk. paper)
    1. Systems engineering--Mathematical models. I. Junkins, John L. II. Title. III. Series.

TA168.S525 2008
620.001'171--dc22   2008026812

Visit the Taylor & Francis Web site at
http://www.taylorandfrancis.com

and the CRC Press Web site at
http://www.crcpress.com

To my parents and my wife Shaweta for their love and constant encouragement.

Puneet Singla

To my wife Elouise for her patience and support.

John L. Junkins

Contents

Preface

Most engineers, scientists and applied mathematicians are familiar with approximation methods and utilize them to varying degrees, either as the focus of their work or as a tool to solve problems. A large set of specialized approximation methods has evolved rapidly over recent decades; the research literature on these methods has evolved in a rather fragmented fashion. For example, the literature on finite element methods is being advanced mainly by researchers addressing aspects of computational mechanics with a focus on structures, materials, thermal sciences or fluid mechanics, and virtually all of the work and resulting software address problems in spatial dimensions of three or less. Likewise, a substantial fraction of the literature and software on artificial neural networks is being published by analysts studying input/output mapping, pattern recognition, or inference in either computer science or control systems. Similarly, much of the work on wavelets is being done by analysts motivated by two-dimensional image processing problems. The level of abstraction and mathematical rigor varies widely across various subsets of the literature. There is also a significant applied mathematics approximation theory literature that cuts across boundaries of the engineering methodology domain.

The discipline-focused decomposition of approximation methods is the natural outgrowth of the "technological pull" associated with the needs for improved solutions of specific classes of problems. The dark side of research literature decomposition into relatively weakly coupled subsets is that the methods are frequently being pursued in a notation-disjointed fashion, published in specialized journals, without reference to a common theoretical foundation. As a consequence, pursuit of generalized methods taking advantage of cross-fertilization opportunities is difficult. Nonetheless, very substantial progress is being made, and there are many opportunities for acceleration and generalization. These opportunities are being driven by a number of forces, including the nanotechnology revolution, wherein we need multi-resolution modeling methods to span nine orders of magnitude of length and time scales and to accommodate atomic/molecular models on the nano scale and ultimately map into continuum models on the macro scale. Another important set of drivers is the emergence of needs to solve important partial differential equations governing nonlinear dynamic state space flows and diffusion in N dimensions (such as the Fokker-Planck and Hamilton-Jacobi-Bellman equations). The latter needs cry out for generalized methods that are optimized with regard to the vector-valued curse of dimensionality.

This text is written from a particular philosophical point of view that will emerge as the chapters unfold. The key ingredient is to develop an "open stance" with regard to understanding available approaches (or developing new ones). Doing research in approximation methods does not require one to adopt a particular "religious faith" in (choose one): finite difference methods, finite volume methods, "traditional" finite element methods, meshless finite element methods, radial basis functions, artificial neural methods, wavelets or some new method. The reality is that many analysts frequently develop expertise in using only one or two approaches and their faith in what they understand colors, perhaps unconsciously, their vision and diminishes their ability to entertain other approaches that may be advantageous. A more difficult set of concerns arises when an analyst learns to use a computer code (based on one particular approach) in a "black box fashion," without coming to appreciate important nuances and restrictions implicit in the coded algorithms that affect the validity of the results produced.

Against this backdrop, we have engaged over the past few years in research that seeks to unify and generalize some of the most important methodology in a new way. The four main aspects of the approach we pursue throughout this text are the following:

- Develop underlying approximation theory from first principles in a fashion accessible by senior undergraduates and especially first-year graduate students in engineering and applied science, i.e., build a rigorous foundation upon which modern approximation methods can be formulated broadly.

- Blend classical methods from probability theory and estimation theory with the methods from approximation theory to place approximation methods in a common theoretical framework.

- Present traditional and novel methods in a common notational framework; apply multiple methods to solve a set of benchmark problems. We use comparisons of competing solutions of these benchmark problems with the goal of providing a qualitative appreciation of the relative and absolute utility of the several approaches. In general, we seek to help the reader develop an evidence-based perspective on making "which method is best" judgments.

- Develop generalizations of existing methodology with special emphasis on extending available methods to solve multi-resolution approximation problems in an N-dimensional space. We use these generalized tools to solve benchmark problems that provide a basis for practical assessments of "what does it all mean?"

The development of several ideas and algorithms is pursued with a variable level of abstraction in the formulation and discussion. The intent is to permit qualitative understanding to be easily achieved, but with attention to developing the most important ideas with sufficient rigor and generality. The use of benchmark problems throughout the presentation, illustrating some of the computational implications, is an important feature of this text that should accelerate understanding and increase appreciation of the material.

Writing this book, while challenging at times, really has been a lot of fun. We trust that you will gain significant insights when reading our presentation and come to appreciate the new concepts and philosophical approach we have taken. Because this field is changing, and also like any human endeavor, we have fallen short of our ambitious goals, so we are confident that this book is not the final word and that many readers will have occasion to significantly extend the concepts presented. We hope that readers are able to evolve these ideas to continue advancement of the field. Most important, we hope you find these ideas useful to solve specific problems that arise in your work.

Puneet Singla
John L. Junkins

Acknowledgments

We are indebted to many colleagues and organizations that have contributed directly to the developments documented in this text or have indirectly contributed through discussions and support of our effort over the past few years. We are delighted to recognize these contributions in five groups, as follows.

First, we express our sincere appreciation for the contributions of the following colleagues, listed in alphabetical order:

Satya Atluri, Harold Black, Christian Bruccoleri, Suman Chakravorty, Goong Chen, John Crassidis, Remi Engles, Troy Henderson, John Hurtado, Michael Jacox, James Jancaitis, Jer-Nan Juang, Anup Katake, Bong Su Ko, Mrinal Kumar, Andrew Kurdila, Dimitris Lagoudas, Manoranjan Majii, Gary Miller, Daniele Mortari, Othon Rediniotis, Lawrence Robertson, Rush Robinett, Tarunraj Singh, Theofanis Strouboulis, Kamesh Subbarao, James Turner, Srinivas Vadali, John Valasek, and Lisa Willingham. Your contributions are warmly appreciated, and where appropriate, your work is referenced herein. We give a special thank you to Andrew Kurdila who read two drafts of this manuscript and made valuable suggestions and also to Lisa Willingham who has supported our effort in many indirect ways.

Second, we owe much to the support, over the years, provided by our academic homes at the University of Virginia, Virginia Polytechnic Institute and State University, Texas A&M University, and the University at Buffalo. Thank you to all the leaders and colleagues at these great institutions who contributed to the rich academic atmosphere in which we work.

Third, our governmental and industrial research sponsors have generously underwritten much of this work, frequently indirectly as a part of research projects directed toward particular classes of systems or applications. We mention especially the following: NASA, the Office of Naval Research, the U.S. Air Force Office of Scientific Research, the U.S. Army Office Research Office, the Naval Surface Weapons Center, Sandia National Laboratory, the State of Texas Advanced Technology Research Program, the Johns Hopkins Applied Physics Laboratory, the Naval Surface Weapons Center, McDonnell Douglas Missile and Space Systems (now Boeing), Lockheed Martin Corporation, and Star Vision Technologies, Inc.

Fourth, we express our sincere thanks to the staff at CRC Press, especially to Bob Stern and Amber Donley, who were very patient when we overran deadlines; we trust the final manuscript enhancements justify your patience. Special thanks are due to our book editor Christine Andreasen for her assistance.

Finally, we owe our special thanks to our families, most especially our wives, Shaweta and Elouise, for their patience and support while we wrestled this manuscript into final submission.

1

Least Squares Methods

Life stands before me like an eternal spring with new and brilliant clothes.

C. F. Gauss

1.1 Introduction

In all branches of engineering, various system processes are generally characterized by mathematical models. Controller design, optimization, fault detection, and many other advanced engineering techniques are based upon mathematical models of various system processes. The accuracy of the mathematical models directly affect the accuracy of the system design and/or control performance. As a consequence, there is a great demand for the development of advanced modeling algorithms that can adequately represent the system behavior. However, different system processes have their own unique characteristics which they do not share with other structurally different systems. Obviously the mathematical structure of engineering models are very diverse; they can be simple algebraic models, may involve differential, integral or difference equations or may be a hybrid of these. Further, many different factors, like intended use of the model, problem dimensionality, quality of the measurement data, offline or online learning, etc., can result in ad-hoc decisions leading to an inappropriate model architecture. For the simplest input-output relationship, the mapping from the state to the measurable quantities is approximated adequately by a linear algebraic equation:

$$\bar{Y} = a_1 x_1 + a_2 x_2 + ... + a_n x_n \tag{1.1}$$

where Y and a_i denote the measured variables and x_i denotes the unknown parameters that characterize the system. So the problem reduces to the estimation of the true but unknown parameters (x_i) from certain data measurements. When the approximation implicit in Eq. (1.1) is satisfactory, we have a linear algebraic estimation problem. The problem of linear parameter estimation arises in a variety of engineering and applied science disciplines such as economics, physics, system realization, signal processing, control, parameter identification, etc. The unknown parameters of a system can be constant or time varying depending upon the system characteristics. For instance, the stability coefficients of the Boeing-747 for cruising flight are constant with respect to time while for the space shuttle these parameters change with time over the re-entry trajectory. All estimation algorithms fall into the category of either a Batch Estimator or a Sequential Estimator, depending upon the way in which observation data are processed. A batch estimator simultaneously processes all data in a single "batch" to estimate the optimum state vector while a sequential estimator is based upon a recursive algorithm, which updates the state vector as soon as each subset of new measurements arrive. Typically, the batch of measurements results in many more (m) equations of the form Eq. (1.1) than (n) the number of to-be-estimated parameters (x_i). Due to their recursive nature, the sequential estimators are preferred for real time estimation problems but either can be used for static and dynamic estimation problems. The batch estimator results are generally more sensitive to model errors and may require some kind of post analysis if large model errors exist. In this chapter, the detailed formulation and analysis of classical batch and sequential least squares algorithms are presented.

1.2   The Least Squares Algorithm

In 1801, Gauss developed the least squares method for determining best estimates for the orbits of the inner planets. Amazingly, during this same time frame, Gauss also introduced the Gaussian (normal) probability distribution, the maximum likelihood principle and the beginning of computational linear algebra (he developed Gaussian elimination to solve systems of algebraic equations). He put all of these powerful new tools to immediate use in his foundational work on least squares orbit determination. Algorithms based on Gauss' least squares minimization principle remain the most widely used methods for estimation of the constant state vector from the set of redundant observations, even after two centuries. In least squares estimation, the optimum state vector is obtained by minimizing the sum of squares of the vertical offsets ("residuals") of the points from the best-fitting approximation. The sum of squares of the residuals as a minimization has several theoretical justifications; note that it provides the differentiable continuity to the loss function [1]. While it does indeed have theoretical and mathematical advantages, it may occasionally lead to poor approximations fit depending upon the problem. Qualitatively, the vertical offsets are frequently preferred over the perpendicular offsets (normal to the best fitting approximation) due to following reasons:

1. It is easier to specify the uncertainties in data along the x- or y-axis rather than in the perpendicular direction.

2. The estimator formulation is much easier and computationally more efficient in case of vertical offset.

3. The difference between vertical offset fit and perpendicular offset fit is negligible for a large number of data points and "relatively small" measurement errors.

1.3 Linear Least Squares Methods

In this section, the linear least squares problem is discussed in detail. First we introduce the *Batch Least Squares* algorithm followed by the *Sequential Least Squares* algorithm.

1.3.1 Batch Least Squares Method

As mentioned in Section 1.2, a batch estimator processes the data taken from a fixed time span, so let us assume that $\tilde{\mathbf{y}}$ is an $m \times 1$ vector of the measured values of signal (y_j) at measurement time t_j and $\mathbf{x}$ is an $n \times 1$ vector of constant parameters to be estimated. The simplest mathematical relationship between $\mathbf{y}$ and $\mathbf{x}$ is defined as

$$\mathbf{y} = \mathbf{Hx} \tag{1.2}$$

where $\mathbf{H}$ is an $m \times n$ matrix of specified independent basis functions. Further measurements are modeled as an unknown true value plus some error, i.e., $\tilde{\mathbf{y}} = \mathbf{y} + \nu$. The measurement error can be any kind of sensor error, modeling error of actual process or due to some other unknown reason, but usually measurement errors are modeled as a Gaussian noise process with known covariance $\mathbf{R}$.

$$E(\nu) = 0 \text{ and } E(\nu\nu^T) = \mathbf{R} \tag{1.3}$$

The estimated values of $\mathbf{x}$ are represented by $\hat{\mathbf{x}}$ and the difference between measured value of signal $(\tilde{\mathbf{y}})$ and estimated value of signal $(\hat{\mathbf{y}})$ is known as residual error $(\mathbf{e})$. So finally we have

$$\tilde{\mathbf{y}} = \mathbf{Hx} + \nu \tag{1.4}$$

$$\tilde{\mathbf{y}} = \mathbf{H}\hat{\mathbf{x}} + \mathbf{e} \tag{1.5}$$

As mentioned earlier that optimum least squares solution is obtained by minimizing the vertical offset between measured values and estimated values, and therefore the loss function to be minimized is given by

$$J(\hat{\mathbf{x}}) = \frac{1}{2}\mathbf{e}^T\mathbf{We} = \frac{1}{2}(\tilde{\mathbf{y}} - \mathbf{H}\hat{\mathbf{x}})^T\mathbf{W}(\tilde{\mathbf{y}} - \mathbf{H}\hat{\mathbf{x}}) \tag{1.6}$$

where $\mathbf{W}$ is an $m \times m$ symmetric positive definite weight matrix. In the diagonal case, it consists of the weights assigned to each measurement. Usually, weights are taken inversely proportional to the measurement precision (variance). The optimum estimate $(\hat{\mathbf{x}})$ of constant state vector $(\mathbf{x})$ is found by satisfying the following necessary and sufficient conditions:

Necessary Condition

$$\frac{\partial}{\partial \mathbf{x}} J|_{\hat{\mathbf{x}}} = \mathbf{H}^T \mathbf{W} \tilde{\mathbf{y}} - (\mathbf{H}^T \mathbf{W} \mathbf{H}) \hat{\mathbf{x}} = 0 \tag{1.7}$$

Sufficient Condition

$$\frac{\partial^2}{\partial \mathbf{x} \mathbf{x}^T} J|_{\hat{\mathbf{x}}} = \mathbf{H}^T \mathbf{W} \mathbf{H} \geq 0 \text{ (i.e., must be positive definite)} \tag{1.8}$$

Now, Eq. (1.7) yields the solution for the optimum estimated state vector $(\hat{\mathbf{x}})$ as

$$\hat{\mathbf{x}} = (\mathbf{H}^T \mathbf{W} \mathbf{H})^{-1} \mathbf{H}^T \mathbf{W} \tilde{\mathbf{y}} \tag{1.9}$$

and from Eq. (1.8), we can conclude that the *Hessian* matrix $(\mathbf{H}^T \mathbf{W} \mathbf{H})$ is positive definite if $\mathbf{H}$ has rank n and the weight matrix $(\mathbf{W})$ is positive definite. It can be shown statistically that the optimum choice of weight matrix $\mathbf{W}$ is the inverse of measurement error covariance matrix $\mathbf{R}$, i.e., $\mathbf{W}|_{opt} = \mathbf{R}^{-1}$ [2]. In general, $\mathbf{W}$ is fully populated if the measurement errors are correlated. By this substitution for $\mathbf{W}$ in Eq. (1.9), we obtain an expression for the *Gauss-Markov theorem*:

$$\hat{\mathbf{x}} = (\mathbf{H}^T \mathbf{R}^{-1} \mathbf{H})^{-1} \mathbf{H}^T \mathbf{R}^{-1} \tilde{\mathbf{y}} \tag{1.10}$$

The covariance matrix for the estimate error is defined as the second moment of difference between the estimated state vector and the true state vector:

$$\mathbf{P} = E[(\hat{\mathbf{x}} - \mathbf{x})(\hat{\mathbf{x}} - \mathbf{x})^T] \tag{1.11}$$

Making use of the *parallel axis theorem* [1], we can write

$$\mathbf{P} = E(\hat{\mathbf{x}} \hat{\mathbf{x}}^T) - E(\mathbf{x}) E(\mathbf{x})^T \tag{1.12}$$

The next step in computing the error covariance matrix is to compute the expectation of $\hat{\mathbf{x}} \hat{\mathbf{x}}^T$:

$$E(\hat{\mathbf{x}} \hat{\mathbf{x}}^T) = [(\mathbf{H}^T \mathbf{R}^{-1} \mathbf{H})^{-1} \mathbf{H}^T \mathbf{R}^{-1}] E(\tilde{\mathbf{y}} \tilde{\mathbf{y}}^T) [(\mathbf{H}^T \mathbf{R}^{-1} \mathbf{H})^{-1} \mathbf{H}^T \mathbf{R}^{-1}]^T \tag{1.13}$$

Substituting for $\tilde{\mathbf{y}}$ from Eq. (1.4) and assuming that state vector $(\mathbf{x})$ and measurement noise vector (ν) are uncorrelated, and further using Eq. (1.3), we get

$$E(\hat{\mathbf{x}} \hat{\mathbf{x}}^T) = \mathbf{M}^{-1} \mathbf{H}^T \mathbf{R}^{-1} [\mathbf{R} + \mathbf{H}(E(\mathbf{x} \mathbf{x}^T)) \mathbf{H}^T] \mathbf{R}^{-1} \mathbf{H} (\mathbf{M}^{-1})^T \tag{1.14}$$

where $\mathbf{M} = \mathbf{H}^T \mathbf{R}^{-1} \mathbf{H}$ is a symmetric matrix. Now, simplifying the expression in Eq. (1.14) using the fact that $\mathbf{x}$ is the true deterministic quantity, i.e., $E(\mathbf{x}) = \mathbf{x}$, we obtain

$$E(\hat{\mathbf{x}} \hat{\mathbf{x}}^T) = \mathbf{M}^{-1} + \mathbf{x} \mathbf{x}^T \tag{1.15}$$

and finally, substituting Eq. (1.15) in Eq. (1.12), we obtain an expression for error covariance matrix $\mathbf{P}$:

$$\mathbf{P} = (\mathbf{H}^T \mathbf{R}^{-1} \mathbf{H})^{-1} \tag{1.16}$$

It is noteworthy that the scaling of the weight matrix $\mathbf{W}$ does not affect the least squares solution given by Eq. (1.9) but results in reciprocal scaling of the covariance matrix $\mathbf{P}$ given by Eq. (1.16). This lack of uniqueness of the weight matrix means that we can get the same value of state estimates for different values of weight matrix even though corresponding covariance information is different.

It is natural to speculate on the efficiency of the least squares estimator. The efficiency of an unbiased estimator can be evaluated by using the Cramer-Rao inequality [2]. According to the theory underlying the Cramer-Rao inequality, the lower bound on state error covariance matrix $\mathbf{P}$ is given by Fisher information matrix $\mathcal{F}$:

$$\mathbf{P} = E((\hat{\mathbf{x}} - \mathbf{x})(\hat{\mathbf{x}} - \mathbf{x})^T) \geq \mathcal{F}^{-1} \tag{1.17}$$

where the Fisher information matrix $\mathcal{F}$ is given by the following Hessian matrix [2]:

$$\mathcal{F}^{-1} = E(\frac{\partial}{\partial \mathbf{x}\mathbf{x}^T} \ln f(\tilde{\mathbf{y}}; \mathbf{x})) \tag{1.18}$$

where $f(\tilde{\mathbf{y}}; \mathbf{x})$ is the probability density function of measurement samples $\tilde{\mathbf{y}}$. To compute the Fisher information matrix $\mathcal{F}$ for the least squares estimation problem, we can use the loss function J given by Eq. (1.6) instead of the probability density function $f(\tilde{\mathbf{y}}; \mathbf{x})$, i.e.,

$$\mathcal{F} = E(\frac{\partial}{\partial \mathbf{x}\mathbf{x}^T} J) \tag{1.19}$$

From Eqs. (1.19), (1.44) and (1.17), we get

$$\mathbf{P} \geq (\mathbf{H}^T \mathbf{W} \mathbf{H})^{-1} \tag{1.20}$$

From Eq. (1.16), the equality in Eq. (1.20) is satisfied and therefore we can conclude that the least squares estimator is the most efficient possible estimator.

1.3.2 Sequential Least Squares Algorithm

Most estimation problems arising in the motion of vehicles (on the ground, in air or space, or underwater) are real time estimation problems, i.e., the unknown state parameters need to be estimated quickly and continuously over the time span of the problem. The demand for efficient real time recursive estimation is frequently acute for high speed aerospace vehicles. Therefore, we need a recursive estimation algorithm which can estimate the unknown state vector in a near continuous manner. As mentioned in the previous section the least squares algorithm is the most efficient possible estimator, so we would like to convert the batch least squares algorithm into a *sequential least squares algorithm*. This is accomplished by deriving a sequential filter based upon using new measurement subsets to update previous state estimates.

Let us consider two measurement subsets $\tilde{\mathbf{y}}_1$ and $\tilde{\mathbf{y}}_2$ at times t_1 and t_2, respectively.

$$\tilde{\mathbf{y}}_1 = \mathbf{H}_1 \mathbf{x} + \nu_1 \tag{1.21}$$

$$\tilde{\mathbf{y}}_2 = \mathbf{H}_2 \mathbf{x} + \nu_2 \tag{1.22}$$

The optimum least squares estimate $\hat{\mathbf{x}}$ of the unknown state vector $\mathbf{x}$ at times t_1 and t_2 is given by Eq. (1.10):

$$\hat{\mathbf{x}}_1 = (\mathbf{H}_1^T \mathbf{R}_1^{-1} \mathbf{H}_1)^{-1} \mathbf{H}_1^T \mathbf{R}_1^{-1} \tilde{\mathbf{y}}_1 \tag{1.23}$$

$$\hat{\mathbf{x}}_2 = (\mathbf{H}^T \mathbf{R}^{-1} \mathbf{H})^{-1} \mathbf{H}^T \mathbf{R}^{-1} \tilde{\mathbf{y}} \tag{1.24}$$

where $\mathbf{R}_1$ and $\mathbf{R}$ are the noise covariance matrices associated with ν_1 and ν, respectively, and $\tilde{\mathbf{y}}$, $\mathbf{H}$ and ν are the merged measurement vector, sensitivity matrix and noise vector at time t_2, respectively.

$$\tilde{\mathbf{y}} = \begin{bmatrix} \tilde{\mathbf{y}}_1 \\ \cdots \\ \tilde{\mathbf{y}}_2 \end{bmatrix}, \mathbf{H} = \begin{bmatrix} \mathbf{H}_1 \\ \cdots \\ \mathbf{H}_2 \end{bmatrix} \text{ and } \nu = \begin{bmatrix} \nu_1 \\ \cdots \\ \nu_2 \end{bmatrix} \tag{1.25}$$

We further assume a block diagonal form for measurement noise vector $\mathbf{R}$, i.e.,

$$\mathbf{R} = \begin{bmatrix} \mathbf{R}_1 & \vdots & \mathbf{O} \\ \cdots & & \cdots \\ \mathbf{O} & \vdots & \mathbf{R}_2 \end{bmatrix} \tag{1.26}$$

The state error covariance matrices for the estimated state vector at times t_1 and t_2 follow from Eq. (1.16):

$$\mathbf{P}_1 = (\mathbf{H}_1^T \mathbf{R}_1^{-1} \mathbf{H}_1)^{-1} \tag{1.27}$$

$$\mathbf{P}_2 = (\mathbf{H}^T \mathbf{R}^{-1} \mathbf{H})^{-1} = (\mathbf{H}_1^T \mathbf{R}_1^{-1} \mathbf{H}_1 + \mathbf{H}_2^T \mathbf{R}_2^{-1} \mathbf{H}_2)^{-1} \tag{1.28}$$

From Eqs. (1.27) and (1.28), we can immediately infer that

$$\mathbf{P}_2^{-1} = \mathbf{P}_1^{-1} + \mathbf{H}_2^T \mathbf{R}_2^{-1} \mathbf{H}_2 \tag{1.29}$$

Now, using Eqs. (1.26), (1.27) and (1.28) we can rewrite Eqs. (1.23) and (1.24) as

$$\hat{\mathbf{x}}_1 = \mathbf{P}_1 \mathbf{H}_1^T \mathbf{R}_1^{-1} \tilde{\mathbf{y}}_1 \tag{1.30}$$

$$\hat{\mathbf{x}}_2 = \mathbf{P}_2 (\mathbf{H}_1^T \mathbf{R}_1^{-1} \tilde{\mathbf{y}}_1 + \mathbf{H}_2^T \mathbf{R}_2^{-1} \tilde{\mathbf{y}}_2) \tag{1.31}$$

Pre-multiplying Eq. (1.30) by $\mathbf{P}_1^{-1}$ and using Eq. (1.29), we can show that

$$(\mathbf{P}_2^{-1} - \mathbf{H}_2^T\mathbf{R}_2^{-1}\mathbf{H}_2)\hat{\mathbf{x}}_1 = \mathbf{H}_1^T\mathbf{R}_1^{-1}\tilde{\mathbf{y}}_1 \tag{1.32}$$

Now, substituting Eq. (1.32) into Eq. (1.31), we get

$$\hat{\mathbf{x}}_2 = \hat{\mathbf{x}}_1 + \mathbf{P}_2(\mathbf{H}_2^T\mathbf{R}_2^{-1}\tilde{\mathbf{y}}_2 - \mathbf{H}_2^T\mathbf{R}_2^{-1}\mathbf{H}_2\hat{\mathbf{x}}_1) \tag{1.33}$$

Finally, rearranging the terms in Eq. (1.33) and defining $\mathbf{K} = \mathbf{P}_2\mathbf{H}_2^T\mathbf{R}_2^{-1}$, we get the expression for a sequential least squares estimator:

$$\hat{\mathbf{x}}_2 = \hat{\mathbf{x}}_1 + \mathbf{K}_2(\tilde{\mathbf{y}}_2 - \mathbf{H}_2\hat{\mathbf{x}}_1) \tag{1.34}$$

Now, we can generalize Eq. (1.34) for the estimation of the unknown state vector $\hat{\mathbf{x}}_{k+1}$ at time t_{k+1} using the previous estimate of state vector $\hat{\mathbf{x}}_k$ and a new measurement set $\tilde{\mathbf{y}}_{k+1}$:

$$\hat{\mathbf{x}}_{k+1} = \hat{\mathbf{x}}_k + \mathbf{K}_{k+1}(\tilde{\mathbf{y}}_{k+1} - \mathbf{H}_{k+1}\hat{\mathbf{x}}_k) \tag{1.35}$$

This equation is known as the *Kalman update equation*, where $\mathbf{K}_{k+1}$ is the $n \times m$ *Kalman gain matrix*, and is the simplest special case version (for linear algebraic systems) of the Kalman filter:

$$\mathbf{K}_{k+1} = \mathbf{P}_{k+1}\mathbf{H}_{k+1}^T\mathbf{R}_{k+1}^{-1} \tag{1.36}$$

$$\mathbf{P}_{k+1}^{-1} = \mathbf{P}_k^{-1} + \mathbf{H}_{k+1}^T\mathbf{R}_{k+1}^{-1}\mathbf{H}_{k+1} \tag{1.37}$$

To compute the unknown state vector recursively using the *Kalman update equation* in the above written form, we have to take the inverse of two $n \times n$ matrices and an $m \times m$ matrix. But using the *matrix inversion lemma* [3] we can rewrite Eq. (1.37) as

$$\mathbf{P}_{k+1} = \mathbf{P}_k - \mathbf{P}_k\mathbf{H}_k^T(\mathbf{H}_{k+1}\mathbf{P}_k\mathbf{H}_{k+1}^T + \mathbf{R}_{k+1})^{-1}\mathbf{H}_{k+1}\mathbf{P}_k \tag{1.38}$$

Now substituting Eq. (1.38) in Eq. (1.36) and further simplifying the expression in the resulting equation, we get

$$\mathbf{K}_{k+1} = \mathbf{P}_k\mathbf{H}_{k+1}^T(\mathbf{H}_{k+1}\mathbf{P}_k\mathbf{H}_{k+1}^T + \mathbf{R}_{k+1})^{-1} \tag{1.39}$$

$$\mathbf{P}_{k+1} = (\mathbf{I} - \mathbf{K}_{k+1}\mathbf{H}_{k+1})\mathbf{P}_k \tag{1.40}$$

The resultant update expression for gain matrix $\mathbf{K}$ and $\mathbf{P}$ in Eqs. (1.39) and (1.40) is computationally more efficient than their earlier counterpart in Eqs. (1.36) and (1.37) as now we are taking the inverse of only one $m \times m$ matrix.

In the sequential least squares estimation algorithm, we update our previous state estimate by an additional correction term given by the *Kalman update equation* using the information contained in the measurement subset at time t_{k+1}. The main problem with the sequential estimator is that it requires an a priori estimate of state vector and the corresponding state error covariance matrix for the initialization process. In the absence of an a priori estimate the batch least squares estimates from the first measurement data set can be used for initialization. In the next chapter, we will show that the *Kalman update equation* plays an important role in the derivation of the classical Kalman filter for dynamical systems whose time evolution is described by differential equations.

1.4   Non-Linear Least Squares Algorithm

In the previous section, we discussed linear least squares algorithms, but unfortunately most of the estimation problems in real world are non-linear. Generally, non-linear estimation problems are solved in an iterative manner by successively approximating the non-linear model by some local linear model.

Let us consider the following non-linear model:

$$\tilde{\mathbf{y}} = \mathbf{h}(\mathbf{x}) + \nu \tag{1.41}$$

In this model, it is assumed that $\mathbf{h}(\mathbf{x})$ is a continuously differentiable function of the state vector $\mathbf{x}$ only and measurement noise, ν, appears only as a separate additive term. As shown previously let us minimize the weighted sum squares vertical offset error, given by

$$J(\mathbf{x}) = \frac{1}{2}[\tilde{\mathbf{y}} - \mathbf{h}(\mathbf{x})]^T \mathbf{W}[\tilde{\mathbf{y}} - \mathbf{h}(\mathbf{x})] \tag{1.42}$$

The optimum estimate of state vector, $\hat{\mathbf{x}}$, can be obtained by satisfying the following necessary and sufficient conditions:

Necessary Condition
$$\frac{\partial}{\partial \mathbf{x}} J|_{\hat{\mathbf{x}}} = 0 \tag{1.43}$$

Sufficient Condition
$$\frac{\partial^2}{\partial \mathbf{x} \mathbf{x}^T} J|_{\hat{\mathbf{x}}} > 0, \text{ (i.e., must be positive definite)} \tag{1.44}$$

However, the non-linear function $\mathbf{h}(\mathbf{x})$ makes the least squares solution difficult to find explicitly from the above generally nonlinear necessary and sufficient conditions. Also, in general, we have no guarantee that a unique solution exists. Gauss developed a widely useful iterative algorithm that usually converges efficiently to the desired solution, given a good initial guess (any point within the domain of attraction of the global minimum, if it exists).

Let us assume that the initial estimate, $\hat{\mathbf{x}}^\star$, of state vector $\mathbf{x}$ is known. Now, we can linearize Eq. (1.41) about $\hat{\mathbf{x}}^\star$:

$$\tilde{\mathbf{y}} - \mathbf{h}(\hat{\mathbf{x}}^\star) = \mathbf{H}(\hat{\mathbf{x}}^\star)\Delta\hat{\mathbf{x}} + \nu \qquad (1.45)$$

where $\mathbf{H}(\hat{\mathbf{x}}^\star) = \frac{\partial \mathbf{h}}{\partial \mathbf{x}}|_{\hat{\mathbf{x}}^\star}$ is an $m \times n$ sensitivity matrix and $\Delta\hat{\mathbf{x}} = \mathbf{x} - \hat{\mathbf{x}}^\star$ is the correction to be applied to our initial estimate ($\hat{\mathbf{x}}^\star$). Now, Eq. (1.45) represents a linear model. Therefore, from Eq. (1.10) we can write an expression for the "differential correction" term ($\Delta\hat{\mathbf{x}}$) as

$$\Delta\hat{\mathbf{x}} = (\mathbf{H}^T(\hat{\mathbf{x}}^\star)\mathbf{W}\mathbf{H}(\hat{\mathbf{x}}^\star))^{-1}\mathbf{H}^T(\hat{\mathbf{x}}^\star)\mathbf{W}(\tilde{\mathbf{y}} - \mathbf{h}(\hat{\mathbf{x}}^\star)) \qquad (1.46)$$

Eq. (1.46) lays a foundation for the Gauss iterative algorithm to find the best possible state estimate. It is an $(m) \times (n)$ over-determined generalization of the classical Newton's root solving method, which of course reduces to Newton's method for $m = n = 1$. The flowchart for the algorithm is shown in Fig. 1.1. The procedure for updating the current estimate continues until

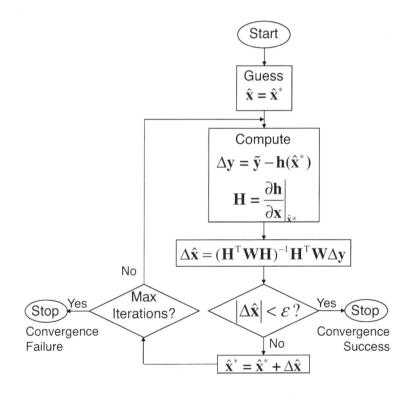

FIGURE 1.1
Non-Linear Least Squares Estimation Algorithm.

$\Delta\hat{\mathbf{x}}$ converges to zero. Convergence difficulties can arise, especially for highly nonlinear systems if the initial guess for state vector $\mathbf{x}$ is far from true value. Besides this, numerical difficulties may arise in the computation of $\Delta\hat{\mathbf{x}}$ as particular problems may have a locally rank-deficient sensitivity matrix, $\mathbf{H}$. Physically/geometrically, this means that sensitivities of the measurements with respect to some variables may be approaching zero, or some rows or columns may not be linearly independent.

1.5   Properties of Least Squares Algorithms

Least squares solutions are generally characterized by the following properties:

1. The least squares estimates are a linear function of the measurement vector $\tilde{\mathbf{y}}$.

2. The least squares estimator is an unbiased estimator. That means the expected value (in the sense of probability theory) of the estimates are free of any kind of systematic error or bias, i.e., $E(\hat{\mathbf{x}}) = E(\mathbf{x}) = \mathbf{x}$.

3. The weight matrix in Eq. (1.9) is not unique. The scaling of all elements of weight matrix $\mathbf{W}$ by a constant parameter (say α) does not affect the least squares solution but scales the covariance matrix by the factor of $\frac{1}{\alpha}$.

4. The residual error ($\mathbf{e}$) of the least squares solution is orthogonal to the range space $\mathcal{R}(\mathbf{H})$ of the sensitivity matrix $\mathbf{H}$, i.e., $\mathbf{H}^T(\tilde{\mathbf{y}} - \mathbf{H}\hat{\mathbf{x}}) = 0$ [2], [1]. This property can be easily proved from Eq. (1.9). For the unity weight matrix $\mathbf{W}$, we can write from Eq. (1.9)

$$\mathbf{H}^T\mathbf{e} = \mathbf{H}^T\left[\tilde{\mathbf{y}} - \mathbf{H}\hat{\mathbf{x}}\right] = \mathbf{H}^T\left[\tilde{\mathbf{y}} - \mathbf{H}\left(\mathbf{H}^T\mathbf{H}\right)^{-1}\mathbf{H}^T\tilde{\mathbf{y}}\right] = \mathbf{H}^T\tilde{\mathbf{y}} - \mathbf{H}^T\tilde{\mathbf{y}} = 0$$

$$(1.47)$$

However, we can violate this property in the case of non-linear estimation.

5. The state estimate error covariance matrix is independent of the measurement data. So we can evaluate the performance of the estimator for different measurement models without taking any actual measurements.

The major advantage of least squares algorithms over other static parameter estimation algorithms is that they are usually easier to design and implement. However, the classical least squares solutions are based on the assumption that the sensitivity matrix ($\mathbf{H}$) is always free of any kind of error. Unfortunately, this assumption is not valid for many practical problems. The inaccuracies in the sensitivity matrix can arise due to modeling or instrumental error. Beside this, the convergence of sequential and non-linear least squares algorithms cannot always be guaranteed. However, all these problems can usually be overcome by a careful analysis or modification of the formulation.

1.6 Examples

1.6.1 Smooth Function Approximation

In this section, we consider the problem of approximating the given input-output data by a continuous function over a compact interval $[a, b]$ using the least squares method. We assume that we have m pairs of observations $(x_1, y_1), \cdots, (x_m, y_m)$ of (x, y), and that a suitable model for the generation of the data is

$$y_i = f(x_i) + \nu_i, \quad i = 1, \cdots, m \tag{1.48}$$

where ν_i denotes random white noise and $f(x)$ is a deterministic one variable continuous function given as follows:

$$f(x) = 1.1(1 - x - 2x^2)e^{-\frac{x^2}{2}} \tag{1.49}$$

We assume that the approximation of $f(x)$, $\hat{f}(x)$ can be written as a linear combination of any prescribed set of linearly independent continuous basis functions $\Phi = \{\phi_i\}_{i=1}^n$:

$$\hat{f}(x) = \sum_{i=0}^n c_i \phi_i(x) \tag{1.50}$$

For illustration purposes, we generate measurement data points in accordance with Eq. (1.48) while taking 5,000 uniform samples of x_i in the interval $[-5, 5]$ and assuming ν_i to be a random variable with zero mean normal distribution and standard deviation equal to 0.01. Now, using this observation data set and polynomial basis functions, i.e., $\phi_i(x) = x^i$, we find the least squares estimates for unknown polynomial coefficients c_i, $i = 0, 1, 2, \cdots, n$. To study the effect of the number of basis functions (order of polynomials) on the approximation accuracy, we vary the order of polynomials from 3 to 8. Fig. 1.2(a) shows the plot of true signal and approximated signals using various orders of polynomials. Further, Figs. 1.2(b) and 1.2(c) show the plots of mean and standard deviation of approximation error. As expected the mean of the approximation error is independent of the order of polynomials used which is in accordance with the fact that the least squares method is an unbiased estimator. Also, it is clear that as the order of polynomials increases, the standard deviation of approximation error decreases, i.e., approximation accuracy increases. However, one cannot arbitrarily increase the order of polynomials because after some value of n ($\approx \geq 10$) the least squares solution becomes numerically unstable. Thus, we have a paradox: it can be proven that any smooth function can be theoretically approximated to any tolerance by a sufficiently high degree polynomial. We find in practice that the most straightforward algorithm limits the practical approximation error order to a small double digit number. We will see that there are several avenues around this paradox: *(i) we can subdivide the input range and fit locally supported polynomials to achieve high accuracy, (ii) we can fit the data using orthogonal polynomials and avoid the matrix inversion altogether, or (iii) we can use non-polynomial approaches to achieve high precision.* We will discuss this particular problem in much more detail in the next chapter.

1.6.2   Star Camera Calibration

For many important applications, the model we need to fit data is some general nonlinear function derived from the theoretical development of the problem at hand. In these problems, the main issue is frequently the "observability" of the system parameters buried in the nonlinear model. To illustrate these ideas, we consider a practical problem that arises in satellite attitude determination using star cameras.

Star tracker cameras and vision-based sensors are primarily used to determine a spacecraft's attitude and position. *However, no sensor is perfect!* In order to achieve high precision information from these sensors, those systematic affects which tend to introduce error in the information must be accounted for. These effects can include lens distortion and instrument aging. A lot of learning algorithms have been presented in the literature to learn the focal plane distortion map. A detailed overview of calibration of CCD cameras (digital cameras) can be found in Refs. [4,5]. These papers provide a description of the various distortion mechanisms and review means to account for these distortion mechanisms.

The first step in the calibration process is to hypothesize an observation model for the vision sensor. This is usually based on the physical insight regarding the particular sensor. For camera-like sensors, the following collinearity equations are used to model the projection from object space to image space as a function of the attitude of the object:

$$x_i = -f \frac{C_{11}r_{x_i} + C_{12}r_{y_i} + C_{13}r_{z_i}}{C_{31}r_{x_i} + C_{32}r_{y_i} + C_{33}r_{z_i}} + x_0, \ i = 1, 2, \cdots, N \qquad (1.51)$$

$$y_i = -f \frac{C_{21}r_{x_i} + C_{22}r_{y_i} + C_{23}r_{z_i}}{C_{31}r_{x_i} + C_{32}r_{y_i} + C_{33}r_{z_i}} + y_0, \ i = 1, 2, \cdots, N \qquad (1.52)$$

where C_{ij} are the unknown elements of attitude matrix $\mathbf{C}$ associated to the orientation of the image plane with respect to some reference plane, f is the known focal length, (x_i, y_i) are the known image space measurements for the ith line of sight, $(r_{x_i}, r_{y_i}, r_{z_i})$ are the known inertial frame direction components of the ith line of sight and N is the total number of measurements. x_0

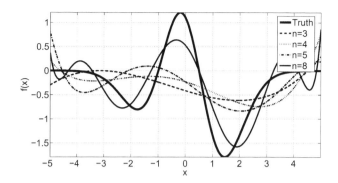

(a) Truth and Approximated Signal

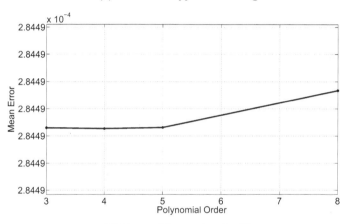

(b) Mean of Approximation Error

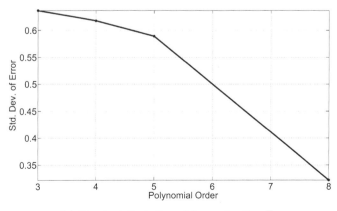

(c) Standard Deviation of Approximation Error

FIGURE 1.2
Least Squares Approximation of Smooth Function of Eq. (1.49).

and y_0 refer to the principal point offset. Generally, the focal plane calibration process is divided into two major parts:

1. Calibration of principal point offset (x_0, y_0) and focal length (f).

2. Calibration of the non-ideal focal plane image distortions due to all other effects (lens distortions, misalignment, detector alignment, etc.).

The implicit pin-hole camera model is not exact. As the first step we need to find the best effective estimates of principal point offset (x_0, y_0) and focal length (f). The residual input-output errors can be captured as departures from this idealized model, and simply lumped into the distortion calibration process. However, the principal point offset is obviously correlated with the inertial pointing of the boresight. In this section, we demonstrate the application of least squares methods discussed in the previous section to learn principal point offset along with the focal length of the camera in a way that does not simultaneously attempt to estimate the sensor pointing. We will make use of interstar angle measurement error which does not depend upon sensor pointing. While this approach leads to reduced observability of (x_0, y_0), we find redundant measurements are sufficient to determine good estimates for (x_0, y_0) and f.

Generally, one needs attitude matrix information for camera calibration; however, we develop an attitude independent approach by making use of the fact that the angle between two star vectors is invariant, whether measured in the image frame or inertial frame. Thus we can form measurement equations that do not depend upon the attitude but instead depend upon the uncertain focal length and principal point offset [6]. There is sufficient information in these measurement equations to estimate the principal point offset and focal length; then we can subsequently obtain the good attitude. Since the interstar angles for the imaged vectors and the inertial frame vectors have to be same, then

$$\cos \theta_{ij} = \mathbf{r}_i^T \mathbf{r}_j = \mathbf{b}_i^T \mathbf{b}_j \qquad (1.53)$$

where $\mathbf{b}_i$ is the star direction vector in the sensor frame corresponding to i^{th} line-of-sight measurement and is given by

$$\mathbf{b}_j = \frac{1}{\sqrt{(x_j - x_0)^2 + (y_j - y_0)^2 + f^2}} \left\{ \begin{array}{c} -(x_j - x_0) \\ -(y_j - y_0) \\ f \end{array} \right\} \qquad (1.54)$$

Substitution of Eq. (1.54) in Eq. (1.53) leads to the following expression:

$$\cos \theta_{ij} = \mathbf{r}_i^T \mathbf{r}_j = \frac{N}{D_1 D_2} = g_{ij}(x_0, y_0, f) \qquad (1.55)$$

where

$$N = (x_i - x_0)(x_j - x_0) + (y_i - y_0)(y_j - y_0) + f^2 \tag{1.56}$$

$$D_1 = \sqrt{(x_i - x_0)^2 + (y_i - y_0)^2 + f^2} \tag{1.57}$$

$$D_2 = \sqrt{(x_j - x_0)^2 + (y_j - y_0)^2 + f^2} \tag{1.58}$$

$$\tag{1.59}$$

Also, notice that Eq. (1.55) equates the cataloged value of $\cos\theta_{ij}$ to the theoretical model which is a nonlinear function of x_0, y_0 and f and therefore we will need to linearize the system about the current estimates of x_0, y_0 and f. Linearization of Eq. (1.55) gives

$$R_{ij} = \mathbf{r}_i^T \mathbf{r}_j - g_{ij}(x_0, y_0, f) = \left[\frac{\partial g_{ij}}{\partial x_0} \frac{\partial g_{ij}}{\partial y_0} \frac{\partial g_{ij}}{\partial f} \right]_{\hat{x}_0, \hat{y}_0, f} \begin{Bmatrix} \delta x_0 \\ \delta y_0 \\ \delta f \end{Bmatrix} \tag{1.60}$$

$$R_{ij} = \mathbf{H}\delta Z \tag{1.61}$$

where $\mathbf{H}$ is the sensitivity matrix and $\delta Z = \left\{ \delta x_0 \; \delta y_0 \; \delta f \right\}^T$. The various derivatives in Eq. (1.60) are given by

$$\frac{\partial g_{ij}}{\partial x_0} = \frac{D_1 D_2 (2x_0 - x_i - x_j) + N[(x_i - x_0)\frac{D_2}{D_1} + (x_j - x_0)\frac{D_1}{D_2}]}{(D_1 D_2)^2} \tag{1.62}$$

$$\frac{\partial g_{ij}}{\partial y_0} = \frac{D_1 D_2 (2y_0 - y_i - y_j) + N[(y_i - y_0)\frac{D_2}{D_1} + (y_j - y_0)\frac{D_1}{D_2}]}{(D_1 D_2)^2} \tag{1.63}$$

$$\frac{\partial g_{ij}}{\partial f} = \frac{D_1 D_2 (2f) + Nf[\frac{D_2}{D_1} + \frac{D_1}{D_2}]}{(D_1 D_2)^2} \tag{1.64}$$

Now, one can use the *nonlinear least squares method* to estimate principal point offset along with focal length of the camera.

For illustration purposes, the simulated spacecraft is assumed to be in a geosynchronous orbit. An $8^0 \times 8^0$ FOV camera is simulated by using the pin-hole camera model with principal point offset of $x_0 = 0.75$ and $y_0 = 0.25$. The focal length of the camera is assumed to be 64.3 mm. Image data are sampled at 10 Hz and centroiding noise of mean zero and standard deviation of 17μ radians is introduced to the true star data. Figs. 1.3(a), 1.3(b) and 1.3(c) show the plot of nonlinear least squares estimates for x_0, y_0 and f, respectively. From these plots, the nonlinear least squares estimates appear to be very noisy. This is due to the fact that in this particular problem, the sensitivity matrix $\mathbf{H}$ is a function of star centroids (x_i, y_i), which includes some random centroiding measurement error as well as systematic errors that are not known. Each image was used to consider all interstar angles as a "batch"; thus x_0, y_0 and f were estimated from each individual image. Alternatively, we could have considered all images simultaneously as one large "batch," with the resulting estimates of x_0, y_0 and f more precise for this case. As another alternative, we can make use of the Kalman filter to process the images recursively. To filter out the noise in nonlinear least squares estimates, the nonlinear least squares estimates for each image are used as "measurements" for a sequential least squares method to to find the best possible estimate of principal point estimate and focal length while combining many least squares estimates. In this case, the sequential least squares solution converges to high precision, to the same best estimate as for the case when all images are considered simultaneously in one large batch process.

Figs. 1.4(a), 1.4(b) and 1.4(c) show the plot of filtered estimates for x_0, y_0 and f, respectively. These simulation results show that the values of (x_0, y_0) and f converge after 250 seconds. The simulation results also reveal the fact that the focal length convergence is quite robust while x_0 and y_0 are less observable with larger oscillations in their convergence transients. This fact can be supported by the SVD decomposition of matrix $\mathbf{H}$ in Eq. (1.61). The singular values of the sensitivity matrix give an idea of degree of observability of the states. Singular values of matrix $\mathbf{H}$ are given below.

$$\mathbf{S} = \begin{bmatrix} 7.53 \times 10^{-4} & 2.14 \times 10^{-5} & 1.54 \times 10^{-5} \end{bmatrix}^T \qquad (1.65)$$

It is clear that two singular values of matrix $\mathbf{H}$ are an order of magnitude lower than the third one (corresponding to the focal length) and therefore principal point offsets are less observable than the focal length. Furthermore, a variance analysis is performed to check the validation of our estimates. Let us consider two body measurements:

$$\tilde{\mathbf{b}}_i = \mathbf{C}(\mathbf{q}) + \nu_i \qquad (1.66)$$

$$\tilde{\mathbf{b}}_j = \mathbf{C}(\mathbf{q}) + \nu_j \qquad (1.67)$$

Now, let us consider the effective measurement z:

$$z = \tilde{\mathbf{b}}_i^T \tilde{\mathbf{b}}_j = \mathbf{r}_i^T \mathbf{r}_j + \mathbf{r}_i \mathbf{C}^T \nu_j + \mathbf{r}_j^T \mathbf{C}^T \nu_i + \nu_i^T \nu_j \qquad (1.68)$$

Since ν_i and ν_j are uncorrelated and represent the zero mean Gaussian process, therefore

$$E(z) = \mathbf{r}_i^T \mathbf{r}_j \qquad (1.69)$$

Defining the new variable p, we get

$$p = z - E(z) = \mathbf{r}_i \mathbf{C}^T \nu_j + \mathbf{r}_j^T \mathbf{C}^T \nu_i + \nu_i^T \nu_j \qquad (1.70)$$

Now computing $E(p^2)$ under the assumption that $E(\nu \nu^T) = \mathbf{R}$ leads to the following expression:

$$\sigma_p^2 = E(p^2) = \mathbf{r}_i \mathbf{C}^T \mathbf{R}_j \mathbf{C} \mathbf{r}_i + \mathbf{r}_j \mathbf{C}^T \mathbf{R}_i \mathbf{C} \mathbf{r}_j + \mathit{Trace}(\mathbf{R}_i \mathbf{R}_j) \qquad (1.71)$$

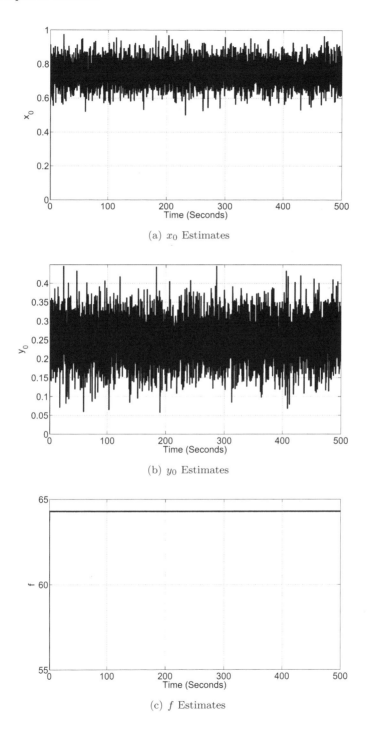

(a) x_0 Estimates

(b) y_0 Estimates

(c) f Estimates

FIGURE 1.3

Nonlinear Least Squares Estimates for Principal Point Offset and Focal Length
of the Camera.

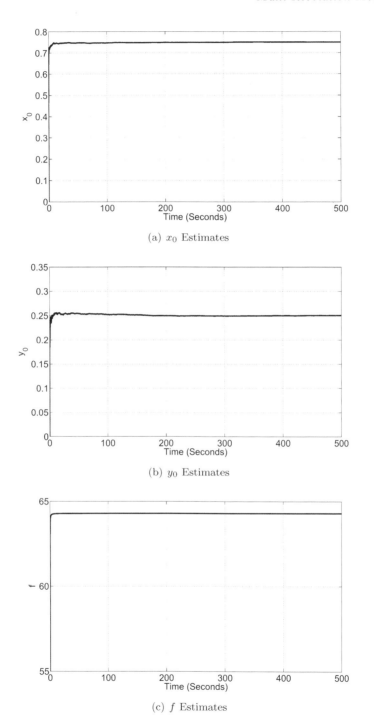

(a) x_0 Estimates

(b) y_0 Estimates

(c) f Estimates

FIGURE 1.4

Sequential Least Squares Estimates for Principal Point Offset and Focal
Length of the Camera.

The last term in Eq. (1.71) is typically of higher order, which can effectively be ignored. If $\mathbf{R}_i = \sigma_i^2 \mathbf{I}$ and $\mathbf{R}_j = \sigma_j^2 \mathbf{I}$ then Eq. (1.71) leads us to the following expression for σ_p^2:

$$\sigma_p^2 = \sigma_i^2 + \sigma_j^2 + 3\sigma_i^2 \sigma_j^2 \tag{1.72}$$

Furthermore, if $\sigma_i^2 = \sigma_j^2 = \sigma^2$ then $\sigma_p^2 = \sigma^2 + 3\sigma^4$. Eq. (1.72) is used to check the consistency of the simulation results and it is verified that the left-hand side of this equation matches well with the right-hand side for the chosen values of the tuning parameters.

1.7 Summary

In this chapter, we briefly discuss classic least squares-based methods to learn unknown parameters of a given mathematical model. These methods will form the basis of various approximation algorithms which will be discussed later in this text. We discuss the utility of these methods by considering two example problems.

2

Polynomial Approximation

We are servants rather than masters in mathematics.

C. Hermite

2.1   Introduction

Basis functions are the fundamental building blocks for most approximation processes. There are infinitely many choices for basis functions such as polynomials, trigonometric functions, radial basis functions, etc. A central difficulty in learning input-output relationships directly from measurements lies in choosing appropriate basis functions and the choice of basis functions unfortunately depends on the characteristics of an unknown input-output map. The problem of choosing an appropriate basis function is difficult since one usually cannot say in advance how complex the input-output map will be, or specify its characteristics. Furthermore, one typically would not have the time or patience to search some handbook of known functions for a set that best represents what we want to study. Hence, we would like to choose building blocks which allow the adaptive construction of input-output maps efficiently and quickly. Fortunately, there is frequently a lack of uniqueness, in that many feasible choices exist for basis functions. So the challenge is not an impossible quest. Consider the Stone-Weierstrass theorem which gives one of the most remarkable results in the field of approximation theory: Qualitatively, there exists a sequence of polynomials that converge uniformly to any prescribed continuous function on a compact interval. This theorem was first stated by Weierstrass for polynomial approximations in 1-D spaces [7] and was later modified by Stone to generalize it for polynomial approximation in compact 2-D spaces [8–10]. For a general compact space, this theorem can be generalized to N dimensions as follows [11].

Stone-Weierstrass Approximation Theorem. *Let X be a compact Hausdorff space and $C[X]$ be a space of continuous functions on X. Then the set of polynomials in N variables form a dense set in $C[X]$.*

As a consequence of this theorem, we can approximate any continuous function on a compact interval with a polynomial of N variables having a sufficient number of terms. The main advantage of using a polynomial basis is that the Fourier coefficients corresponding to each basis function can be computed by solving a system of linear equations governed by the least squares process. This involves the computation of the Vandermonde matrix inverse which is not difficult if the size of linear system is reasonable. As we learned in chapter 1, we should not be overly confident in the Stone-Weierstrass theorem, because the practical degree required for particular cases may introduce numerical difficulties in determining high-order polynomial fits. Fig. 2.1 shows how the condition number (ratio of largest singular value to smallest singular value) of such a matrix increases linearly in the case of polynomial basis functions as the number (or order of polynomial) increases, for the case of a polynomial function of one independent variable. We note that not only does the accuracy degrade due to poor condition number, but also the computational cost to compute a solution with large, fully populated matrices is an issue that deters high-order polynomial approximations. Hence, we will prefer those polynomial basis functions that imply special structures for these equations. In the case of orthogonal basis functions, the Fourier coefficient corresponding to each basis function can be efficiently computed from ratios of inner products, avoiding matrix inversion altogether. Furthermore, Fourier coefficients corresponding to each basis function are independent of each other and so inclusion of any new basis function in the basis vector set does not require us to re-solve for previously computed Fourier coefficients. In this chapter, we discuss in detail two different schemes to construct orthogonal polynomials in both continuous and discrete variables. We first discuss an analogy between a construction scheme (using the hypergeometric differential equation) for continuous variable orthogonal polynomials and discrete variable orthogonal polynomials, followed by approximation properties of the orthogonal polynomials.

2.2   Gram-Schmidt Procedure of Orthogonalization

Let $\mathcal{V}$ be a finite dimensional inner product space spanned by basis vector functions $\{\psi_1(x), \psi_2(x), \cdots, \psi_n(x)\}$. According to the *Gram-Schmidt process* an orthogonal set of basis functions $\{\phi_1(x), \phi_2(x), \cdots, \phi_n(x)\}$ can be constructed from any basis functions $\{\psi_1(x), \psi_2(x), \cdots, \psi_n(x)\}$ by following three steps:

1. Initially there is no constraining condition on the first basis element $\phi_1(x)$ and therefore we can choose $\phi_1(x) = \psi_1(x)$.

2. The second basis vector, orthogonal to the first one, can be constructed

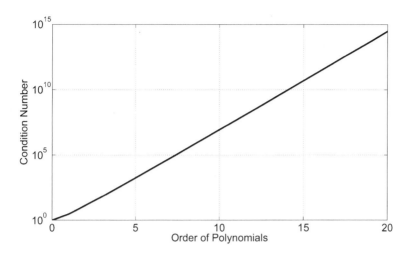

FIGURE 2.1
Condition Number of the Vandermonde Matrix vs. Order of Polynomials.

by satisfying the following condition

$$\langle \phi_2(x), \phi_1(x) \rangle = 0 \tag{2.1}$$

where $\langle f(x), g(x) \rangle$ denotes the inner product between two functions $f(x)$ and $g(x)$. We define the inner product for the continuous case as

$$\langle f(x), g(x) \rangle = \int_{x \in \Omega} w(x) f(x) g(x) dx \tag{2.2}$$

and for the discrete case as

$$\langle f(x), g(x) \rangle = \sum_{i=1}^{N} w(x_i) f(x_i) g(x_i) \tag{2.3}$$

where $w(x) \geq 0$ is a specified weight function. Typically $\{x_1 < x_2 < \cdots < x_N\}$ lie on a closed interval of x-space.

Further, if we write

$$\phi_2(x) = \psi_2(x) - c\phi_1(x) \tag{2.4}$$

then we can determine the following value of unknown scalar constant c by substituting this expression for $\phi_2(x)$ in the orthogonality condition, given by Eq. (2.1):

$$c = \frac{\langle \psi_2(x), \phi_1(x) \rangle}{\langle \phi_1(x), \phi_1(x) \rangle} \tag{2.5}$$

3. Continuing the procedure listed in step 2, we can write $\phi_k(x)$ as

$$\phi_k(x) = \psi_k(x) - c_1\phi_1(x) - c_2\phi_2(x) - \cdots - c_{k-1}\phi_{k-1}(x) \qquad (2.6)$$

where the unknown constants $c_1, c_2, \cdots, c_{k-1}$ can be determined by satisfying the following orthogonality conditions:

$$\langle \phi_k(x), \phi_j(x) \rangle = 0 \text{ for } j = 1, 2, \cdots, k-1 \qquad (2.7)$$

Since $\phi_1(x), \phi_2(x), \cdots, \phi_{k-1}(x)$ are already orthogonal to each other the scalar constant c_j can be written as

$$c_j = \frac{\langle \psi_k(x), \phi_j(x) \rangle}{\langle \phi_j(x), \phi_j(x) \rangle} \qquad (2.8)$$

Therefore, finally we have the following general *Gram-Schmidt formula* for constructing the orthogonal basis vectors $\phi_1(x), \phi_2(x), \cdots, \phi_n(x)$:

$$\phi_k(x) = \psi_k(x) - \sum_{j=1}^{k-1} \frac{\langle \psi_k(x), \phi_j(x) \rangle}{\langle \phi_j(x), \phi_j(x) \rangle} \phi_j(x), \text{ For } k = 1, 2, \cdots, n \qquad (2.9)$$

2.2.1   Three-Term Recurrence Relation to Generate Orthogonal Polynomials

Let $\mathcal{V}_n$ be a finite dimensional inner product space spanned by orthogonal basis vector functions $\{\phi_1(x), \phi_2(x), \cdots, \phi_n(x)\}$, where $\phi_n(x)$ represents a polynomial of degree n. Next, since $x\phi_n(x) \in \mathcal{V}_{n+1}$, therefore, there exist numbers $c_0, c_1, \cdots, c_{n+1}$ such that the following is true:

$$x\phi_n(x) = \sum_{i=0}^{n+1} c_{i,n}\phi_i(x) \qquad (2.10)$$

Since $\phi_0(x), \phi_1(x), \cdots, \phi_n(x)$ are orthogonal to each other with respect to the weight function $w(x)$, we have

$$c_{k,n} = \frac{1}{\mu_k^2} \langle x\phi_n(x), \phi_k(x) \rangle, \ \ k = 0, 1, \cdots, n+1 \qquad (2.11)$$

where $\langle ., . \rangle$ denotes the inner product defined by the weight function $w(x)$ and $\mu_k^2 = \langle \phi_k(x), \phi_k(x) \rangle$. Further, note that for $k \leq n-2$, $x\phi_k(x) \in \mathcal{V}_{n-1}$ and hence $c_{k,n} = 0$, $\forall 0 \leq k \leq n-2$ and Eq. (2.10) reduces to

$$x\phi_n(x) = c_{n-1,n}\phi_{n-1}(x) + c_{n,n}\phi_n(x) + c_{n+1,n}\phi_{n+1}(x) \qquad (2.12)$$

Now let us assume that a_n and b_n are the leading coefficients of basis function $\phi_n(x)$, i.e., $\phi_n(x) = a_nx^n + b_nx^{n-1} + \cdots$. Hence, from Eq. (2.10), we get

$$a_n = c_{n+1,n}a_{n+1}, \ \ b_n = c_{n,n}a_n + c_{n+1,n}b_{n+1} \qquad (2.13)$$

Also, substituting for $k = n - 1$ in Eq. (2.11), we get

$$c_{n-1,n} = \frac{1}{\mu_{n-1}^2} \langle x\phi_n, \phi_{n-1} \rangle = \frac{\mu_n^2}{\mu_{n-1}^2} c_{n,n-1} \qquad (2.14)$$

Now from Eqs. (2.13) and (2.14), we get

$$c_{n+1,n} = \frac{a_n}{a_{n+1}}, \quad c_{n,n} = \frac{b_n}{a_n} - \frac{b_{n+1}}{a_{n+1}}, \quad c_{n-1,n} = \frac{\mu_n^2}{\mu_{n-1}^2} \frac{a_{n-1}}{a_n} \qquad (2.15)$$

Now substituting for various $c_i s$ from Eq. (2.15) in Eq. (2.12), we get the following three-term recurrence relation:

$$x\phi_n(x) = \frac{a_n}{a_{n+1}}\phi_{n+1}(x) + \left(\frac{b_n}{a_n} - \frac{b_{n+1}}{a_{n+1}} \right)\phi_n(x) + \frac{\mu_n^2}{\mu_{n-1}^2} \frac{a_{n-1}}{a_n}\phi_{n-1}(x) \quad (2.16)$$

Finally, from Eq. (2.16), it is clear that given a sequence of numbers $\{a_n\}$ and $\{b_n\}$, one can construct orthogonal polynomials with respect to a given weight function $w(x)$ and any generic inner product. That means the orthogonal polynomial $\phi_n(x)$ is unique up to a normalizing factor. In the following section, we give a more detailed proof of this statement.

2.2.2   Uniqueness of Orthogonal Polynomials

In this subsection, we prove that orthogonal polynomials are unique up to a normalizing factor.

Let $\{\phi_i(x)\}$ and $\{\bar{\phi}_i(x)\}$ be two sets of polynomials which satisfy the following orthogonality condition with respect to a given weight function $w(x)$:

$$\langle \phi_i(x), \phi_j(x) \rangle = \mu_i^2 \delta_{ij} \qquad (2.17)$$
$$\langle \bar{\phi}_i(x), \bar{\phi}_j(x) \rangle = \bar{\mu}_i^2 \delta_{ij} \qquad (2.18)$$

The inner products are over the same Ω (i.e., we assume $\{\phi_i(x)\}$ and $\{\bar{\phi}_i(x)\}$ span the same space). Since $\bar{\phi}_n(x)$ is a polynomial of degree n, therefore, we can write it as a linear combination of polynomials: $\{\phi_0, \phi_1, \cdots, \phi_n\}$

$$\bar{\phi}_n(x) = \sum_{k=1}^{n} c_{k,n} \phi_i(x) \qquad (2.19)$$

Note, by Eq. (2.17), $c_{k,n} = 0$ for $k < n$ and, therefore, $\phi(x)$ and $\bar{\phi}(x)$ should be proportional to each other. However, if the leading coefficient of the polynomial $\phi_n(x)$ is constrained to be one then it is apparent that $\phi_n(x) = \bar{\phi}_n(x)$.

To illustrate the Gram-Schmidt procedure, we seek to construct the orthogonal polynomials of degree $\leq n$ with respect to the weight function, $w(x) = 1 - x$ and inner product defined by Eq. (2.2) on the closed interval $[0, 1]$. Let us consider the non-orthogonal monomial basis $1, x, x^2, \cdots, x^n$.

TABLE 2.1

Continuous Variable Orthogonal Polynomials with
Respect to the Weight Function $w(x) = 1 - x$,
$0 \leq x \leq 1$

Degree	Basis Functions, $\phi_j(x)$
0	1
1	$x - \frac{1}{3}$
2	$x^2 - \frac{4}{5}x + \frac{1}{10}$
3	$x^3 - \frac{9}{7}x^2 + \frac{3}{7}x - \frac{1}{35}$
$\vdots$	$\vdots$
n	$\phi_n(x) = \left[x^n - \sum_{j=0}^{n-1} \frac{<x^n, \phi_j(x)>}{<\phi_j(x), \phi_j(x)>} \phi_j(x) \right]$

Assume that $\phi_0(x), \phi_1(x), \cdots$ denote the resulting orthogonal polynomials.
Now we can begin the process of Gram-Schmidt orthogonalization by letting

$$\phi_0(x) = 1 \tag{2.20}$$

According to Eq. (2.9), the next orthogonal polynomial is

$$\phi_1(x) = x - \frac{\langle x, \phi_0 \rangle}{\langle \phi_0, \phi_0 \rangle} \phi_0(x) = x - \frac{1/6}{1/2} = x - \frac{1}{3} \tag{2.21}$$

Further, recursively using the Gram-Schmidt formula given by Eq. (2.9), we
can generate the orthogonal polynomials given in Table 2.1, including the
recursive form given for $\phi_n(x)$. Fig. 2.2 shows the plot of these polynomials
up to order 3.

We mention, with a suitable definition of the inner product, a variation of
Gram-Schmidt process is valid for discrete valued functions and higher dimensional space. In particular, we use the inner product definition of Eq. (2.3)
to construct the discrete variable orthogonal polynomials with respect to the
weight function, $w(x) = 1 - x$. However, the Gram-Schmidt procedure to
generate discrete variable orthogonal polynomials is a bit tedious since the
various coefficients of the polynomial depend upon the number and distribution of points considered to evaluate the inner product of Eq. (2.3). Table

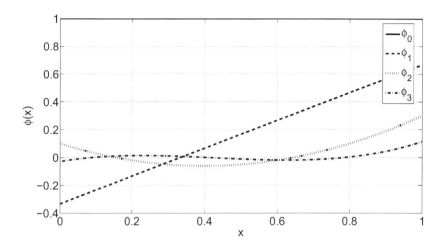

FIGURE 2.2
Continuous Variable Orthogonal Polynomials with Respect to the Weight
Function $w(x) = 1 - x$.

2.2 summarizes the expression for discrete variable orthogonal polynomials
with respect to weight function, $w(x) = 1 - x$ for 100 uniformly and normally
distributed data points between $x = 0$ and $x = 1$. Figs. 2.3(a) and 2.3(b)
show the plot of discrete variable polynomials up to order 3 for uniformly and
normally distributed points, respectively. These plots ascertain the fact that
the discrete variable polynomials can be very different depending upon the
distribution of points. We mention, with forward reference to Chapter 5, that
changing the domain to $-1 \leq x \leq 1$, and $w(x) = 1 - |x|$ gives rise to a related
set of orthogonal polynomials. These polynomials generalize to n dimensions
and are a special case of the Global-Local Orthogonal Mapping (GLO-MAP)
approach to approximation.

Example 2.1
Let us consider the problem of approximating a function with the help of both
conventional and orthogonal polynomial basis functions. We assume that we
have 5,000 pairs of observations $(x_1, y_1), (x_2, y_2)), \cdots, (x_{5000}, y_{5000})$ of (x, y)
in accordance with the following model:

$$y_i = c_0 + \sum_{k=1}^{n} c_k x^k + \nu_i, \quad i = 1, \cdots, m, \ x_i \in [-1, 1] \qquad (2.22)$$

where ν_i denotes measurement noise modeled by the zero mean Gaussian

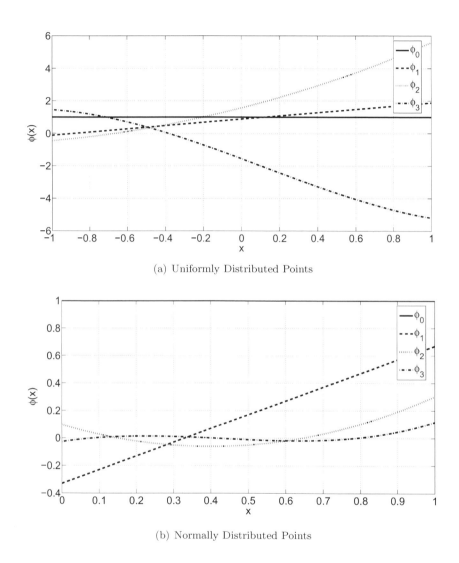

(a) Uniformly Distributed Points

(b) Normally Distributed Points

FIGURE 2.3
Discrete Variable Orthogonal Polynomials with Respect to the Weight Function $w(x) = 1 - x$.

TABLE 2.2
Discrete Variable Orthogonal Polynomials with Respect to the Weight
Function $w(x) = 1 - x$, $0 \le x \le 1$

Degree	Basis Functions (Uniform Distribution of Points), $\phi_j(x)$	Basis Functions (Normal Distribution of Points), $\phi_j(x)$
0	1	1
1	$x - \frac{98}{297}$	$x + \frac{706810}{802991}$
2	$x^2 - \frac{131}{165} + \frac{4753}{49005}$	$x^2 + \frac{1186247}{393260}x + \frac{1222899}{780377}$
3	$x^3 - \frac{295}{231}x^2 + \frac{28814}{68607}x - \frac{20421}{759944}$	$x^3 - \frac{310827}{978016}x^2 - \frac{4060435}{939154}x - \frac{2408782}{1546997}$
$\vdots$	$\vdots$	$\vdots$
n	$x^n - \sum\limits_{j=0}^{n-1} \frac{<x^n, \phi_j(x)>}{<\phi_j(x), \phi_j(x)>}\phi_j(x)$	$x^n - \sum\limits_{j=0}^{n-1} \frac{<x^n, \phi_j(x)>}{<\phi_j(x), \phi_j(x)>}\phi_j(x)$

process of known standard deviation, σ. For simulation purposes, we consider
the following two cases:

1. Perfect Measurements ($\sigma = 0$)

2. Noisy Measurements ($\sigma = 0.01$)

For each of these cases, true measurement signal is generated by varying n
from 5 to 25 in intervals of 5 which gives us five sets of measurement data
points for each case. For each set of these data points, the true coefficients c_i
were chosen randomly, and the 5,000 perfect measurements computed. Furthermore, we approximate these five sets of measurement data points using
linearly independent sets of basis functions:

$$\hat{y}_i = \sum_{k=0}^{n} \hat{c}_k \phi_k(x_i) \qquad (2.23)$$

In the case of conventional polynomials $\phi_i = x^i$ and discrete variables orthogonal polynomial basis functions were generated by using the Gram-Schmidt
orthogonalization process with $w(x) = 1$. For each measurement data point
set, we assumed that we know the orders of polynomials to be used. For example, for the first set of measurement points, we used polynomials up to order

5, and for the fifth measurement data set, we used polynomial basis functions up to order 25. Since we are using polynomial basis functions to approximate data points generated by sampling an appropriate order polynomial function, hence, we expect our approximation errors to be on the order of measurement noise, i.e., $\sigma = 0$ for Case 1 and $\sigma = 0.01$ for Case 2. Figs. 2.4(a) and 2.5(a) show the plot of standard deviation of approximation error versus order of polynomial basis functions using both conventional and orthogonal polynomial basis functions for Case 1 and Case 2, respectively. From this plot, it is clear that approximation error is essentially independent of the order of polynomials in the case of orthogonal polynomial basis functions; however, approximation error deteriorates as the order of polynomials increases in the case of conventional polynomial functions, with the degradation becoming increasingly significant above the tenth order. This deterioration in approximation error can be attributed to the fact that the Vandermonde matrix becomes more ill-conditioned as the order of polynomials is increased. This fact is even more apparent in Figs. 2.4(b) and 2.5(b) which show the plot for the norm of state covariance matrix $\mathbf{P}$ versus the order of polynomial basis functions for Case 1 and Case 2, respectively. As expected, the norm of $\mathbf{P}$ explodes in the case of conventional polynomial basis functions whereas in the case of orthogonal polynomials it is more stable. Obviously, in Figs. 2.4(b) and 2.5(b) $\|\mathbf{P}\|$ plots are identical, because we used the same measurement error covariance matrix $\mathbf{R} = \mathbf{W}^{-1} = 10^{-2}\mathbf{I}$.   ⬚

This particular example once again reinforces our earlier observation regarding the numerical stability of orthogonal basis functions. In the following sections, we seek a generic method to generate continuous and discrete variable orthogonal polynomials with respect to a given weight function $w(x)$ and to obtain useful insight regarding many properties of orthogonal polynomial basis functions.

2.3   Hypergeometric Function Approach to Generate Orthogonal Polynomials

In this section, a more efficient and generic method based upon hypergeometric function theory is described for the construction of continuous and discrete variable orthogonal polynomials with respect to a given weight function. The new method of constructing orthogonal polynomials is based upon the fact that continuous variable orthogonal polynomials satisfy a hypergeometric differential equation while discrete variable polynomials satisfy a corresponding hypergeometric difference equation. For example, classic n^{th} order Legendre

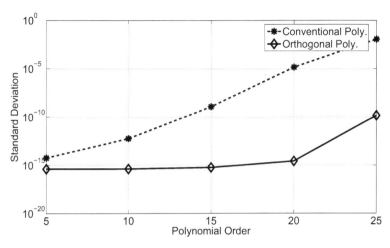

(a) Standard Deviation of Approximation Error

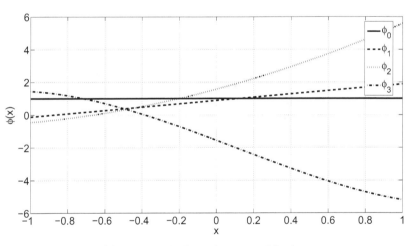

(b) Norm of the State Covariance Matrix

FIGURE 2.4
Case 1: Perfect Measurements.

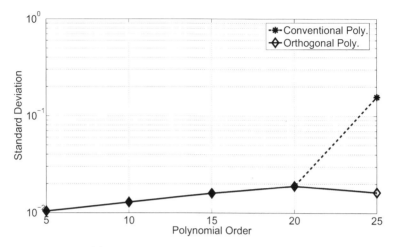

(a) Standard Deviation of Approximation Error

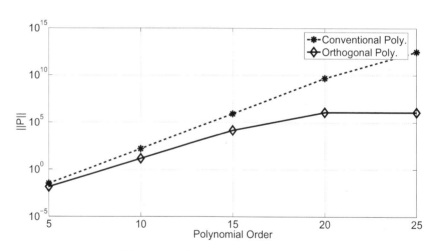

(b) Norm of the State Covariance Matrix

FIGURE 2.5
Case 2: Noisy Measurements ($\sigma = 0.01$).

polynomials, $L_n(x)$, satisfy the following hypergeometric differential equation:

$$(1 - x^2)L_n(x)'' - 2xL_n(x)' + n(n + 1)L_n(x) = 0 \tag{2.24}$$

In general, continuous variable orthogonal polynomials satisfy a generalized hypergeometric differential equation of the following form:

$$\sigma(x)\phi''(x) + \tau(x)\phi'(x) + \lambda\phi(x) = 0 \tag{2.25}$$

where $\sigma(x)$ and $\tau(x)$ are polynomials of at most second and first degree, respectively, and λ is a constant.

To prove the aforementioned fact, let us differentiate Eq. (2.25) with respect to x and then, substituting $P_1 = \phi'(x)$, we get

$$\sigma(x)P_1''(x) + (\sigma'(x) + \tau(x))\,P_1'(x) + (\tau'(x) + \lambda)\,P_1(x) = 0 \tag{2.26}$$

Since $\tau_1 \triangleq \sigma'(x) + \tau(x)$ is a polynomial of at most degree one and $\lambda_1 \triangleq \tau'(x) + \lambda$ is a constant, it is clear that the first derivative of ϕ satisfies the differential equation of the following form:

$$\sigma(x)P_1''(x) + \tau_1(x)P_1'(x) + \lambda_1 P_1(x) = 0 \tag{2.27}$$

In general, the nth derivative of $\phi(x)$, $P_n(x) \triangleq \phi^{(n)}(x)$ satisfies the hypergeometric equation

$$\sigma(x)P_n(x)'' + \tau_n(x)P_n'(x) + \lambda_n P_n(x) = 0 \tag{2.28}$$

where

$$\tau_n(x) = \tau(x) + n\sigma'(x) \tag{2.29}$$

$$\lambda_n = \lambda + n\tau'(x) + \frac{1}{2}n(n - 1)\sigma''(x) \tag{2.30}$$

Now note that if $\lambda_n = 0$, then $P_n(x) = const$ is the solution of Eq. (2.28), and we conclude that $\phi(x) = \phi_n(x)$ is a polynomial of degree n. Hence, for $\lambda = -n\tau'(x) - \frac{1}{2}n(n - 1)\sigma''(x)$, one can construct a family of polynomials which satisfies the hypergeometric differential equation of Eq. (2.25).

Further, to show the orthogonality of this family of polynomials, let us consider a weight function $w(x)$ such that the following is true:

$$(\sigma(x)w(x))' = \tau(x)w(x) \tag{2.31}$$

Multiplying Eq. (2.25) by the weight function $w(x)$ leads to

$$(\sigma(x)w(x)\phi'(x))' + \lambda w(x)\phi(x) = 0 \tag{2.32}$$

Now let us consider hypergeometric differential equations for $\phi_n(x)$ and $\phi_m(x)$ in the following form:

$$(\sigma(x)w(x)\phi_n'(x))' + \lambda_1 w(x)\phi_n(x) = 0 \tag{2.33}$$

$$(\sigma(x)w(x)\phi_m'(x))' + \lambda_2 w(x)\phi_m(x) = 0 \tag{2.34}$$

where

$$\lambda_1 = -n\tau'(x) - \frac{1}{2}n(n-1)\sigma''(x) \qquad (2.35)$$

$$\lambda_2 = -m\tau'(x) - \frac{1}{2}m(m-1)\sigma''(x) \qquad (2.36)$$

Multiplying Eqs. (2.33) and (2.34) by ϕ_m and ϕ_n, respectively, and further, subtracting the Eq. (2.34) from Eq. (2.33), we get

$$\phi_m(x)(\sigma(x)w(x)\phi_n'(x))' - \phi_n(x)(\sigma(x)w(x)\phi_m'(x))' = (\lambda_2 - \lambda_1)w(x)\phi_m(x)\phi_n(x) \qquad (2.37)$$

Now let us consider the terms on L.H.S of the above equation:

$$\phi_m(x)[\sigma(x)w(x)\phi_n'(x)]' - \phi_n(x)[\sigma(x)w(x)\phi_m'(x)]' = \cdots$$
$$\sigma(x)w(x)[\phi_n''(x)\phi_m(x) - \phi_m''(x)\phi_n(x)] + (\sigma(x)w(x))'[\phi_n'(x)\phi_m(x)\cdots$$
$$-\phi_m'(x)\phi_n(x)] = [\sigma(x)w(x)Y(\phi_m(x), \phi_n(x))]' \qquad (2.38)$$

where $Y(\phi_m(x), \phi_n(x)) = \phi_n'(x)\phi_m(x) - \phi_m'(x)\phi_n(x)$.

Now integrating Eq. (2.37) using Eq. (2.38), we get

$$(\lambda_2 - \lambda_1)\int_a^b w(x)\phi_m(x)\phi_n(x)dx = (\lambda_2 - \lambda_1)\langle\phi_m(x), \phi_n(x)\rangle$$

$$= \sigma(x)w(x)Y(\phi_m(x), \phi_n(x))|_a^b \quad (2.39)$$

Hence, if the weight function, $w(x)$, is chosen in such a way that the R.H.S. of the above-written equation is zero and $\lambda_1 \neq \lambda_2$, then polynomial solutions $\phi_m(x)$ and $\phi_n(x)$ are orthogonal to each other with respect to the weight function $w(x)$. Hence, two different solutions of Eq. (2.25) will be orthogonal to each other with respect to some weight function $w(x)$ on interval $[a, b]$ if the following two conditions are satisfied:

$$\sigma(a)w(a) = \sigma(b)w(b) = 0 \qquad (2.40)$$

$$\tau'(x) + \frac{1}{2}(n+m-1)\sigma''(x) \neq 0 \qquad (2.41)$$

Note that the orthogonality requirement of Eq. (2.40) is satisfied if the weight function, $w(x)$, is zero at the end points of the interval $[a, b]$, i.e., the weight function is compactly supported on the interval $[a, b]$.

2.3.1   Derivation of Rodrigues's Formula for Continuous Variable Polynomials

In this section, we derive an expression for the classical Rodrigues's formula for the orthogonal polynomial solution of Eq. (2.25) which gives the fundamental relationship between the nth degree polynomial $\phi_n(x)$ and its nth derivative.

Let us consider a generic weight function w_n such that the following is true:

$$(\sigma(x)w_n(x))' = \tau_n(x)w_n(x) \tag{2.42}$$

Note that the aforementioned expression is a generalization of the expression of Eq. (2.31). Now multiplying Eq. (2.28) by the weight function $w(x)$ leads to

$$(\sigma(x)w_n(x)P_n'(x))' + \lambda_n w_n(x)P_n(x) = 0 \tag{2.43}$$

Substituting for $\tau_n(x)$ from Eq. (2.29) in Eq. (2.42) yields

$$\frac{(\sigma(x)w_n(x))'}{w_n(x)} = \tau(x) + n\sigma'(x) = \frac{(\sigma(x)w(x))'}{w(x)} + n\sigma'(x) \tag{2.44}$$

Further simplification of Eq. (2.44) yields the following relation between $w_n(x)$ and $w(x)$.

$$\frac{(\sigma(x)w_n(x))'}{w_n(x)} = \frac{w'(x)}{w(x)} + n\frac{\sigma'(x)}{\sigma(x)} \tag{2.45}$$

From Eq. (2.45), it follows that

$$w_n(x) = \sigma^n(x)w(x) \tag{2.46}$$

As a consequence of this relationship and the fact that $P_{n+1}(x) = P_n'(x)$, Eq. (2.43) can be rewritten as

$$w_n(x)P_n(x) = -\frac{1}{\lambda_n}(\underbrace{\sigma(x)w_n(x)}_{w_{n+1}(x)}\,P_n'(x))' = -\frac{1}{\lambda_n}\left(w_{n+1}(x)P_{n+1}(x)\right)$$

Now for $n = 0$, Eq. (2.47) can be written as

$$w_0(x)P_0(x) = w(x)\phi(x) = -\frac{1}{\lambda_1}\left(w_1(x)P_1(x)\right) \tag{2.47}$$

Recursively making use of Eq. (2.47), we have

$$w(x)\phi(x) = \left(-\frac{1}{\lambda_1}\right)\left(-\frac{1}{\lambda_2}\right)\left(w_2(x)P_2(x)\right) = \cdots = \frac{1}{A_n}\left(w_n(x)P_n(x)\right)^{(n)} \tag{2.48}$$

where

$$A_n = (-1)^n \prod_{k=0}^{n-1} \lambda_k \tag{2.49}$$

Further, if $\phi(x) = \phi_n(x)$ is assumed to be a polynomial of degree n, then $P_n(x) = \phi_n^{(n)}$ is a constant. Making use of this fact and Eq. (2.46), we get the following relationship between $\phi_n(x)$ and its nth derivative $P_n(x)$:

$$\phi_n(x) = \frac{\phi_n^{(n)}}{A_n w(x)}\left[\sigma^n(x)w(x)\right]^{(n)} = \frac{B_n}{w(x)}\left[\sigma^n(x)w(x)\right]^{(n)} \tag{2.50}$$

where $B_n = \phi_n^{(n)} A_n^{-1}$ and A_n is given by Eq. (2.49) with

$$\lambda_k = \lambda + k\tau'(x) + \frac{1}{2}k(k-1)\sigma''(x), \ \ k = 0, 1, \cdots \tag{2.51}$$

$$\lambda = -n\tau'(x) - \frac{1}{2}n(n-1)\sigma''(x) \tag{2.52}$$

The aforementioned relationship of Eq. (2.50) is known as *Rodrigues's formula* which gives the equation for a generic orthogonal polynomial solution for continuous hypergeometric Eq. (2.25).

2.3.2   Leading Coefficients for Three-Term Recurrence Formula

As discussed earlier, given a sequence of leading coefficients $\{a_n\}$ and $\{b_n\}$, one can construct the orthogonal polynomials with respect to a given weight function $w(x)$ using the three-term recurrence relationship of Eq. (2.16). In this section, we derive a generic expression for leading coefficients $\{a_n\}$ and $\{b_n\}$ while making use of the classical Rodrigues's formula.

Let us assume that an nth order orthogonal polynomial can be written as

$$\phi_n(x) = a_n x^n + b_n x^{n-1} + \cdots$$

and differentiating it $n - 1$ times yields

$$\phi_n^{(n-1)}(x) = n!a_n x + (n-1)!b_n \tag{2.53}$$

It should be noted that $\phi_n^{(n-1)}(x) = P_{n-1}(x)$ is a polynomial of degree one and satisfies the hypergeometric differential Eq. (2.28). Hence, making use of Rodrigues's formula for $\phi_n^{(n-1)}(x)$, we have

$$\phi_n^{(n-1)}(x) = \frac{A_{n-1}B_n}{\sigma^{n-1}(x)w(x)} \frac{d}{dx}\left[\sigma^n(x)w(x)\right] \tag{2.54}$$

where

$$A_{n-1} = (-1)^{n-1} \prod_{k=0}^{n-2} \lambda_k \tag{2.55}$$

$$\lambda_k = -(n-k)\left(\tau'(x) + \frac{n+k-1}{2}\sigma''(x)\right) \tag{2.56}$$

Comparing Eqs. (2.53) and (2.54), we have

$$n!a_n x + (n-1)!b_n = \frac{A_{n-1}B_n}{\sigma^{n-1}(x)w(x)} \frac{d}{dx}\left[\sigma^n(x)w(x)\right] \tag{2.57}$$

Once again, making use of the fact that $(\sigma(x)w_n(x))' = \tau_n(x)w_n(x)$, we get

$$n!a_nx + (n-1)!b_n = \frac{A_{n-1}B_n}{w_{n-1}(x)}\underbrace{\frac{d}{dx}[\sigma(x)w_{n-1}(x)]}_{\tau_{n-1}(x)w_{n-1}(x)} = A_{n-1}B_n\tau_{n-1}(x)$$

$$= A_{n-1}B_n\left(\tau_{n-1}(0) + \tau'_{n-1}x\right) \qquad (2.58)$$

Now by comparing the coefficients of constant and linear terms in x, we get the following relationships for leading coefficients a_n and b_n:

$$a_n = \frac{A_{n-1}B_n}{n!}\tau'_{n-1}(x) \qquad (2.59)$$

$$\frac{b_n}{a_n} = n\frac{\tau_{n-1}(0)}{\tau'_{n-1}(x)} \qquad (2.60)$$

Hence, one can easily construct the orthogonal polynomials to a given weight function $w(x)$ by making use of the above identities and formulating the corresponding hypergeometric differential equation. To illustrate the whole procedure, let us construct a hypergeometric differential equation for the continuous variable orthogonal polynomials shown in Table 2.1. Let us assume the following forms for $\sigma(x)$ and $\tau(x)$ while making use of the fact that $\sigma(x)$ and $\tau(x)$ are polynomials of second and first order, respectively.

$$\sigma(x) = ax^2 + bx + c \qquad (2.61)$$

$$\tau(x) = dx + e \qquad (2.62)$$

Once we know $\sigma(x)$ and $\tau(x)$, the constant λ can be constructed using the following identity:

$$\lambda = -n\tau'(x) - \frac{1}{2}n(n-1)\sigma''(x) \qquad (2.63)$$

Now we need to find unknown coefficients a, b, c, d and e such that the following constraints are satisfied:

$$(\sigma(x)w(x))' = \tau(x)w(x), \forall x, \ w(x) = 1 - x \qquad (2.64)$$

$$\sigma(0)w(0) = \sigma(1)w(1) = 0 \qquad (2.65)$$

Substituting for $\sigma(x)$ and $\tau(x)$ from Eqs. (2.61) and (2.62) in Eqs. (2.64) and (2.65), we get

$$(-3a+d)x^2 + (e-d-2b+2a)x + b - e = 0, \forall x \qquad (2.66)$$

$$c = 0 \qquad (2.67)$$

Since Eq. (2.66) is valid for all values of x, by equating the coefficients of different powers of x to zero, we get

$$3a = d \qquad (2.68)$$

$$e - d = 2(b - a) \qquad (2.69)$$

$$b = e \qquad (2.70)$$

Solving the above system of three linear equations for (a, d, e) as a function of b yields the following values for a, b, c, d and e:

$$a = -b, \ e = b, \ c = 0, \ d = -3b \qquad (2.71)$$

Substituting for these values in Eqs. (2.61) and (2.62), we get the following expressions for $\sigma(x)$ and $\tau(x)$:

$$\sigma(x) = b(-x^2 + 1) \qquad (2.72)$$
$$\tau(x) = b(-3x + 1) \qquad (2.73)$$

Further, from Eq. (2.63), we get the following expression for λ:

$$\lambda = \left(2n + n^2\right) b \qquad (2.74)$$

Hence, the family of orthogonal polynomials w.r.t. weight function, $w(x) = 1 - x$, is the solution of the following hypergeometric differential equation:

$$b(-x^2 + 1)\phi''(x) + b(-3x + 1)\phi'(x) + b\left(2n + n^2\right)\phi(x) = 0 \qquad (2.75)$$

Since this equation is valid for all values of x, we can assume $b = 1$ without any loss of generality:

$$(-x^2 + 1)\phi''(x) + (-3x + 1)\phi'(x) + \left(2n + n^2\right)\phi(x) = 0 \qquad (2.76)$$

Now using the expression for $\tau_{n-1}(x)$ of Eq. (2.29), we have

$$\tau_{n-1}(x) = (-1 - 2\,n)\,x + n \qquad (2.77)$$

which yields the following expression for leading coefficients:

$$\frac{b_n}{a_n} = \frac{n^2}{-1 - 2\,n} \qquad (2.78)$$

It is easy to check that the orthogonal polynomials of Table 2.1 satisfy the hypergeometric differential equation of Eq. (2.76) and their leading coefficients are in accordance with the relationship given by Eq. (2.78).

2.4 Discrete Variable Orthogonal Polynomials

In the previous section, a procedure based upon hypergeometric function theory is discussed in detail to generate continuous variable orthogonal polynomials with respect to any given weight function. The main advantage of this approach is that it provides us analytical formulas for generic expression of orthogonal polynomials and is more convenient than the Gram-Schmidt orthogonalization process. In this section, we seek to extend this methodology for the case of discrete variable orthogonal polynomials. We will briefly discuss how discrete variable orthogonal polynomials can be constructed from the solution of a hypergeometric difference equation which can be obtained directly from the corresponding continuous case hypergeometric differential equation. For more details, readers should refer to Ref. [12].

2.4.1 Hypergeometric Type Difference Equation

In the case of a discrete independent variable, the orthogonality condition can be written as

$$\sum_i p_n(x_i)w(x_i)p_m(x_i) = d_{mn}^2 \delta_{mn} \tag{2.79}$$

Since the expression for discrete variable orthogonal polynomials depends upon the distribution of points x_i, we will assume uniform distribution of points $(x_{i+1} = x_i + h)$ for hypergeometric difference equation development in this section.

From the previous section, we recall that all continuous variable orthogonal polynomials are solutions of the following hypergeometric differential equation:

$$\sigma_c(x)\phi'' + \tau_c(x)\phi'(x) + \lambda_c\phi(x) = 0 \tag{2.80}$$

We will approximate this differential equation by a difference equation assuming a uniform mesh of size $\delta x = h$ on input space:

$$\sigma_c(x)\frac{1}{h}\left[\frac{\phi(x+h)-\phi(x)}{h} - \frac{\phi(x)-\phi(x-h)}{h}\right] + \frac{\tau_c(x)}{2}\left[\frac{\phi(x+h)-\phi(x)}{h}\right.$$
$$\left. +\frac{\phi(x)-\phi(x-h)}{h}\right] + \lambda_c\phi(x) = 0 \tag{2.81}$$

Now a linear change of independent variable x by hx_1 yields

$$\frac{\sigma_c(hx_1)}{h^2}\left[\phi(h(x_1+1)) - 2\phi(hx_1) - \phi(h(x_1-1))\right] + \frac{\tau_c(hx_1)}{2h}\left[\phi(h(x_1+1))\right.$$
$$\left. -\phi(h(x_1-1))\right] + \lambda_c\phi(hx_1) = 0 \tag{2.82}$$

Further, let us assume that $h = 1$ and define the following two difference operators which are equivalent to the differentiation operator in continuous time:

$$\text{Forward: } \triangle f(x) = f(x+1) - f(x) \tag{2.83}$$
$$\text{Backward: } \nabla f(x) = f(x) - f(x-1) \tag{2.84}$$

Now using Eqs. (2.83) and (2.84), we can rewrite Eq. (2.82) as

$$\sigma_c(x_1)\triangle\nabla\phi(x_1) + \frac{\tau_c(x_1)}{2}\left[\triangle + \nabla\right]\phi(x_1) + \lambda_c\phi(x_1) = 0 \tag{2.85}$$

From the definitions of these difference operators, $\triangle$ and ∇, one can easily derive the following identities:

$$\triangle f(x) = \nabla f(x+1) \tag{2.86}$$

$$\nabla f(x) = \triangle f(x) - \nabla \triangle f(x) \tag{2.87}$$

$$\triangle \nabla f(x) = \nabla \triangle f(x) \tag{2.88}$$

$$\triangle [f(x)g(x)] = f(x)\triangle g(x) + g(x+1)\triangle f(x) \tag{2.89}$$

Now using the above-written identity of Eq. (2.87) in Eq. (2.85), we obtain the following difference equation:

$$\sigma(x)\triangle\nabla\phi(x) + \tau(x)\triangle\phi(x) + \lambda\phi(x) = 0 \tag{2.90}$$

where

$$\sigma(x) = \sigma_c(x) - \frac{1}{2}\tau_c(x) \tag{2.91}$$

$$\tau(x) = \tau_c(x) \tag{2.92}$$

$$\lambda = \lambda_c \tag{2.93}$$

Now, analogous to the development in the previous section, we will show that the discrete variable orthogonal polynomials provide a solution of the hypergeometric difference Eq. (2.90). Note that analogous to their continuous counterparts, $\sigma(x)$ and $\tau(x)$ are at most second order and first order polynomials, respectively. To prove the aforementioned fact, let us apply the difference operator $\triangle$ to both sides of Eq. (2.90) and then substitute $P_1(x) = \triangle\phi(x)$:

$$\sigma(x)\triangle\nabla P_1(x) + (\tau(x) + \triangle\sigma(x))\triangle P_1(x) + (\lambda + \triangle\tau(x))P_1(x) = 0 \tag{2.94}$$

Since $\tau_1(x) \triangleq \tau(x) + \triangle\sigma(x)$ is a polynomial of at most degree one and $\lambda_1 \triangleq \lambda + \triangle\tau(x)$ is a constant, it is clear that $P_1(x)$ satisfies the difference equation of the following form:

$$\sigma(x)\triangle\nabla P_1(x) + \tau_1(x)\triangle P_1(x) + \lambda_1 P_1(x) = 0 \tag{2.95}$$

In general, $P_m(x) \triangleq \triangle^m \phi(x)$ satisfies the following hypergeometric difference equation:

$$\sigma(x)\triangle\nabla P_m(x) + \tau_m(x)\triangle P_m(x) + \lambda_m P_m(x) = 0 \tag{2.96}$$

where

$$\tau_m(x) = \tau_{m-1}(x+1) + \triangle\sigma(x) \tag{2.97}$$

$$\lambda_m = \lambda_{m-1} + \triangle\tau_{m-1}(x) = \lambda + m\triangle\tau(x) + \frac{1}{2}m(m-1)\triangle^2\sigma(x) \tag{2.98}$$

Once again, from Eq. (2.96), it can be inferred that if $\lambda_m = 0$, then $P_m(x) = const$ is the solution of Eq. (2.96). Hence, for $\lambda = -m\triangle\tau(x) - \frac{1}{2}m(m-$

1)$\triangle^2\sigma(x)$, i.e., $\lambda_m = 0$ one can construct a family of polynomials which satisfies the hypergeometric difference equation of Eq. (2.90). Now we will show that this family of polynomials are orthogonal polynomials with respect to the weight function $w(x)$ such that the following is true:

$$\triangle\left[\sigma(x)w(x)\right] = \tau(x)w(x) \tag{2.99}$$

Note that this constraint on the expression for weight function is similar to its continuous counterpart given by Eq. (2.31). Now multiplying both sides of Eq. (2.90) by $w(x)$ leads to

$$w(x)\sigma(x)\triangle\nabla\phi(x) + w(x)\tau(x)\triangle\phi(x) + \lambda w(x)\phi(x) = 0 \tag{2.100}$$

Further, using identities of Eq. (2.86) and Eq. (2.89) with $f(x) = \sigma(x)w(x)$ and $g(x) = \nabla\phi(x)$, we can rewrite the above equation as

$$w(x)\sigma(x)\triangle\nabla\phi(x) + \triangle\left[\sigma(x)w(x)\right]\nabla\phi(x+1) + \lambda w(x)\phi(x) = 0$$
$$\triangle\left[\sigma(x)w(x)\nabla\phi(x)\right] + \lambda w(x)\phi(x) = 0 \tag{2.101}$$

Now let us consider the hypergeometric difference equation for solutions $\phi_n(x)$ and $\phi_m(x)$ which are polynomials of degree n and m, respectively,

$$\triangle\left[\sigma(x)w(x)\nabla\phi_n(x)\right] + \lambda_1 w(x)\phi_n(x) = 0 \tag{2.102}$$
$$\triangle\left[\sigma(x)w(x)\nabla\phi_m(x)\right] + \lambda_2 w(x)\phi_m(x) = 0 \tag{2.103}$$

where

$$\lambda_1 = -n\triangle\tau(x) - \frac{1}{2}n(n-1)\triangle^2\sigma(x) \tag{2.104}$$

$$\lambda_2 = -m\triangle\tau(x) - \frac{1}{2}m(m-1)\triangle^2\sigma(x) \tag{2.105}$$

Now multiplying Eqs. (2.102) and (2.103) by $\phi_m(x)$ and $\phi_n(x)$, respectively, and subtracting Eq. (2.103) from Eq. (2.102), we get

$$(\lambda_2 - \lambda_1)\,w(x)\phi_m(x)\phi_n(x) = \phi_m(x)\triangle\left[\sigma(x)w(x)\nabla\phi_n(x)\right]$$
$$-\phi_n(x)\triangle\left[\sigma(x)w(x)\nabla\phi_m(x)\right] \tag{2.106}$$

Further, making use of identities of Eq. (2.86) and Eq. (2.89), we have

$$\phi_m(x)\triangle\left[\sigma(x)w(x)\nabla\phi_n(x)\right] = \{\phi_m(x)\triangle\left[\sigma(x)w(x)\nabla\phi_n(x)\right]$$
$$+\triangle\phi_m(x)\sigma(x+1)w(x+1)\triangle\phi_n(x)\}$$
$$= \triangle\left(\sigma(x)w(x)\phi_m(x)\nabla\phi_n(x)\right) \tag{2.107}$$

Similarly,

$$\phi_n(x)\triangle\left[\sigma(x)w(x)\nabla\phi_m(x)\right] = \triangle\left(\sigma(x)w(x)\phi_n(x)\nabla\phi_m(x)\right) \tag{2.108}$$

Finally, substituting for Eqs. (2.107) and (2.108) in Eq. (2.106), we get

$$(\lambda_2 - \lambda_1)\, w(x)\phi_m(x)\phi_n(x) = \triangle \{\sigma(x)w(x)\left[\phi_m(x)\nabla\phi_n(x) - \phi_n(x)\nabla\phi_m(x)\right]\} \tag{2.109}$$

Now if we sum over $x = x_i$ for $x_i \in [a, b]$, we get

$$(\lambda_m - \lambda_n) \sum_{x_i=a}^{b-1} w(x_i)\phi_m(x_i)\phi_n(x_i) = (\sigma(x)w(x)\left[\phi_m\triangle\phi_n - \phi_n\triangle\phi_m\right]) \mid_a^{b+1} \tag{2.110}$$

Hence, analogous to the continuous case, two different solutions of Eq. (2.90) will be orthogonal to each other on interval $[a, b]$, with respect to the given weight function $w(x)$ and the discrete inner product of Eq. (2.3) if the following two conditions are satisfied:

$$\sigma(a)w(a) = \sigma(b+1)w(b+1) = 0 \tag{2.111}$$

$$\triangle\tau(x) + \frac{1}{2}(n + m - 1)\triangle^2\sigma(x) \neq 0 \tag{2.112}$$

However, it would be sufficient if weight function $w(x)$ is compactly supported over the interval $[a, b+1]$, i.e., $w(a) = w(b+1) = 0$. It should be noted that in the continuous case the corresponding condition was $w(a) = w(b) = 0$.

2.4.2   Derivation of Rodrigues's Formula for Discrete Variable Orthogonal Polynomials

In this section, we will derive the expression for the Rodrigues-type formula for the discrete variable orthogonal polynomial solution of Eq. (2.90) which gives the fundamental relationship between the nth degree polynomial $\phi_n(x)$ and its nth derivative $P_n(x) \triangleq \triangle^n\phi(x)$.

Let us consider a generic weight function $w_n(x)$ such that the following is true:

$$\triangle\left[w_n(x)\sigma(x)\right] = \tau_n(x)w_n(x) \tag{2.113}$$

Note that the aforementioned expression is a generalization of the expression of Eq. (2.99). Multiplying Eq. (2.96) by the weight function $w_n(x)$ leads to

$$\triangle\left[w_n(x)\sigma(x)\nabla P_n(x)\right] + \lambda_n w_n(x)P_n(x) = 0 \tag{2.114}$$

Using the definition of $\triangle$ operator, we can rewrite Eq. (2.113) as

$$w_n(x+1)\sigma(x+1) = w_n(x)\left[\sigma(x) + \tau_n(x)\right] \tag{2.115}$$

Further, substituting for $\tau_n(x)$ from Eq. (2.97) in the above equation leads to

$$\frac{w_n(x+1)\sigma(x+1)}{w_n(x)} = [\sigma(x) + \tau_n(x)]$$

$$= [\sigma(x) + \tau_{n-1}(x+1) + \sigma(x+1) - \sigma(x)]$$

$$= [\tau_{n-1}(x+1) + \sigma(x+1)]$$

$$= \frac{w_{n-1}(x+2)\sigma(x+2)}{w_{n-1}(x+1)}$$

which implies the following relationship between $w_n(x)$ and $w(x)$:

$$w_n(x) = \sigma(x+1)w_{n-1}(x+1) = w(x+n)\prod_{k=1}^{n}\sigma(x+k) \qquad (2.116)$$

As a consequence of this relationship, Eq. (2.114) can be rewritten as

$$w_n(x)P_n(x) = -\frac{1}{\lambda_n}\left[\triangle\left\{w_n(x)\sigma(x)\nabla P_n(x)\right\}\right]$$

$$= -\frac{1}{\lambda_n}\left[\nabla\left\{w_n(x+1)\sigma(x+1)\nabla P_n(x+1)\right\}\right]$$

$$= -\frac{1}{\lambda_n}\left[\nabla\left\{w_n(x+1)\sigma(x+1)\triangle P_n(x)\right\}\right] \qquad (2.117)$$

Further, making use of the fact that $P_{n+1} = \triangle P_n$ and Eq. (2.116), we get

$$w_n(x)P_n(x) = -\frac{1}{\lambda_n}\left[\nabla\left\{w_{n+1}(x)P_{n+1}(x)\right\}\right] \qquad (2.118)$$

$$= \left(-\frac{1}{\lambda_m}\right)\left(-\frac{1}{\lambda_{n-1}}\right)\left[\nabla^2\left\{w_{n+2}(x)P_{n+2}(x)\right\}\right] \quad (2.119)$$

$$\vdots \qquad (2.120)$$

$$= \frac{1}{A_n}\nabla^n\left(w_n(x)P_n(x)\right) \qquad (2.121)$$

where $\phi = \phi_n(x)$ is a polynomial of degree n and

$$A_n = (-1)^n\prod_{k=0}^{n-1}\lambda_k, \quad A_0 = 1 \qquad (2.122)$$

where λ_k is given as

$$\lambda_k = (k-n)\triangle\tau(x) + \frac{1}{2}(k-n)(k+n-1)\triangle^2\sigma(x)$$

$$= -(n-k)\left(\triangle\tau(x) + \frac{k+n-1}{2}\triangle^2\sigma(x)\right) \qquad (2.123)$$

It should be noted that the aforementioned relationship of Eq. (2.118) is the discrete analog of Rodrigues's formula for the continuous case.

2.4.3 Leading Coefficients for Three-Term Recurrence Formula for Discrete Variable Orthogonal Polynomials

As discussed earlier, given a sequence of leading coefficients $\{a_n\}$ and $\{b_n\}$, one can construct the orthogonal polynomials with respect to a given weight function $w(x)$ using the three-term recurrence relationship of Eq. (2.16). In this section, we derive a generic expression for leading coefficients $\{a_n\}$ and $\{b_n\}$ while making use of the discrete analog to Rodrigues's formula.

Let us assume that the nth order orthogonal polynomial can be written as

$$\phi_n(x) = a_n x^n + b_n x^{n-1} + \cdots$$

Now let us define $\triangle^{n-1} x^n \triangleq \alpha_n(x + \beta_n)$ which is a linear function of x. Constants α_n and β_n are unknown. Now let us consider the following relationship:

$$\alpha_{n+1}(x + \beta_{n+1}) = \triangle(x^{n+1}) = \triangle^{n-1}[(x+1)^{n+1} - x^{n+1}] \qquad (2.124)$$

Now making use of binomial expansion for $(x+1)^{n+1}$, we have

$$\alpha_{n+1}(x + \beta_{n+1}) = \triangle^{n-1}\left[(n+1)x^n + \frac{(n+1)n}{2}x^{n-1} + \cdots\right]$$

$$= (n+1)\alpha_n(x + \beta_n) + \frac{(n+1)n}{2}\alpha_{n-1} \qquad (2.125)$$

Now equating the coefficients of linear power of x and constant term, we get

$$\alpha_{n+1} = (n+1)\alpha_n \Rightarrow \alpha_n = n! \qquad (2.126)$$

$$\alpha_{n+1}\beta_{n+1} = (n+1)\alpha_n\beta_n + \frac{(n+1)n}{2}\alpha_{n-1} \qquad (2.127)$$

Substitution of Eq. (2.126) in Eq. (2.127) yields the following expression for β_n:

$$\beta_n = \frac{n-1}{2} \qquad (2.128)$$

Hence, we can write

$$\triangle^{n-1}\phi_n(x) = n!a_n\left(x + \frac{n-1}{2}\right) + (n-1)!b_n \qquad (2.129)$$

It should be noted that $\triangle^{n-1}\phi_n(x)$ is a polynomial of degree one and satisfies the hypergeometric differential Eq. (2.96). Hence, making use of Rodrigues's formula for $\triangle^{n-1}\phi_n(x)$, we have

$$\triangle^{n-1}\phi_n(x) = \frac{A_{n-1}B_n}{\sigma^{n-1}(x)w(x)}\triangle\left[\sigma^n(x)w(x)\right] = A_{n-1}B_n\tau_{n-1}(x) \qquad (2.130)$$

where

$$A_{n-1} = (-1)^{n-1}\prod_{k=0}^{n-2}\lambda_k \qquad (2.131)$$

$$\lambda_k = -(n-k)\left(\tau' + \frac{n+k-1}{2}\sigma''\right) \qquad (2.132)$$

Comparing Eqs. (2.129) and (2.130), we have

$$n! a_n \left(x + \frac{n-1}{2} \right) + (n-1)! b_n = A_{n-1} B_n \left(\tau_{n-1}(0) + \tau'_{n-1} x \right) \qquad (2.133)$$

Now by comparing the coefficients of constant and linear terms in x, we get the following relationships for leading coefficients a_n and b_n:

$$a_n = \frac{A_{n-1} B_n}{n!} \tau'_{n-1}(x) \qquad (2.134)$$

$$\frac{b_n}{a_n} = n \frac{\tau_{n-1}(0)}{\tau'_{n-1}(x)} - \frac{1}{2} n(n-1) \qquad (2.135)$$

which are analogous to their continuous counterparts given by Eqs. (2.59) and (2.60). Now making use of the three-term recurrence relationship of Eq. (2.16) and expressions of Eqs. (2.134) and (2.135), one can easily construct the discrete variable orthogonal polynomials to a given weight function $w(x)$ for uniform distribution of the points. The general treatment for the generation of discrete orthogonal polynomials for unevenly spaced x_i points is beyond the scope of this chapter. Readers should refer to Ref. [12] for more details.

2.5 Approximation Properties of Orthogonal Polynomials

Guided by the Weierstrass approximation theorem, polynomial basis functions are a fundamental and attractive choice to approximate continuous functions on a compact space, to within an approximation error ϵ:

$$f(x) = \sum_i a_i \phi_i(x) + \epsilon = \mathbf{a}^T \mathbf{\Phi}(x) + \epsilon \qquad (2.136)$$

where $\mathbf{\Phi}(.)$ is an infinite dimensional vector of linearly independent polynomial functions and $\mathbf{a}$ is a vector of Fourier coefficients corresponding to polynomial functions. However, according to the following theorem, the continuous function $f(.)$ can be approximated by a set of orthogonal polynomials with a countable number of terms instead of infinite terms.

Theorem 1. *Every nontrivial inner product space has an orthonormal polynomial basis and further if $\{\phi_i\}$ is such an orthonormal basis then at most a countable number of Fourier coefficients, $< f, \phi_i >$, are non-zero to approximate a continuous function $f(x)$ to within a prescribed error ϵ. More generally, $\mathbf{\Phi}(.)$ is any complete set of basis functions.*

Proof. Let us define a set $S_n = \{i \in \mathcal{I} : |<f, \phi_i>| > 1/n\}$. Here, $\mathcal{I}$ denotes an uncountable index set and should not be confused with the set of integers. Note, to prove this theorem, one just needs to show that S_n is a finite set. Now if $F = \sum_{j \in S_n} <f, \phi_j> \phi_j$ is the orthogonal projection of f onto the subspace, $\mathcal{U} = span\,[\phi_j : j \in S_n]$ then by the Pythagorean Law

$$\|f\|^2 = \|(f - F) + F\|^2 = \|f - F\|^2 + \|F\|^2 \geq \|F\|^2 = \sum_{j \in S_n} \|<f, \phi_j> \phi_j\|^2.$$

As ϕ_i is an element of the orthonormal basis, i.e., $\|\phi_i\| = 1$, the above expression reduces to

$$\|f\|^2 \geq \sum_{j \in S_n} |<f, \phi_j>|^2 \geq \sum_{j \in S_n} 1/n^2 = card(S_n)/n^2.$$

Now as $\|f\| < \infty$ hence $card(S_n) < \infty$, i.e., S_n is a finite set. $\qquad\square$

According to this theorem, the Fourier coefficients converge to zero as the number of orthogonal polynomial functions approaches infinity. In other words, one needs only a countable number of orthogonal basis functions to approximate a bounded continuous function to a prescribed resolution. In practical terms, we can use a finite series polynomial to locally approximate any given continuous function. To account for the errors introduced due to the truncation of an infinite series polynomial, we state the following theorem which basically gives us a bound for approximation error using any polynomial basis functions up to degree n.

Theorem 2. *Let f be an $n+1$ times differentiable function over the compact interval $[a, b]$, i.e., $f \in \mathcal{C}^{n+1}[a, b]$ and $\hat{f}$ denotes the approximation of the unknown function f using a complete set of polynomial basis functions up to degree n. Further, let x_i, $i = 0, 1, 2, \cdots, n$ be $n+1$ interpolation points in the compact interval $[a, b]$. We have the following approximation error equation:*

$$e \triangleq f(x) - \hat{f}(x) = (x - x_0)(x - x_1) \cdots (x - x_n) \frac{f^{n+1}(\xi(x))}{(n + 1)!} \qquad (2.137)$$

where $\xi(x) = \xi(x_0, x_1, \cdots, x_n)$.

Proof. First, note that if $x = x_i$, $i = 0, 1, 2, \cdots, n$ then the error expression of Eq. (2.137) is trivial. Therefore, we assume that $x \neq x_i$ and define following function

$$F(t) \triangleq f(t) - \hat{f}(t) - \frac{f(x) - \hat{f}(x)}{(x - x_0)(x - x_1) \cdots (x - x_n)}(t - x_0)(t - x_1) \cdots (t - x_n)$$

$$(2.138)$$

Now it is apparent that F has $n + 2$ zeros, namely, $x_0, \cdots, x_n, x$. Now according to Rolle's theorem [11] F^{n+1} has at least one zero, $\xi = \xi(x, x_0, \cdots, x_n)$. This implies that

$$F^{n+1}(\xi(x)) = f^{n+1}(\xi(x)) - \hat{f}^{n+1}(\xi(x))$$

$$- \frac{f(x) - \hat{f}(x)}{(x - x_0)(x - x_1) \cdots (x - x_n)} (n + 1)! = 0$$

and hence we prove that $e(x) = \frac{(x-x_0)(x-x_1)\cdots(x-x_n)}{(n+1)!} f^{n+1}(\xi(x))$ □

Note, in the case of the interpolation problem, the definition of x_i, $i = 1, 2, \cdots, n$ is straightforward and in the case of the least squares approximation, Theorem 3 guarantees the existence of these points, if one uses orthogonal basis functions. Further, if $f^{n+1}(.)$ is bounded by a number M, then Eq. (2.137) can be replaced by the following inequality:

$$e(x) \le \frac{(b - a)^{n+1}}{(n + 1)!} M \tag{2.139}$$

From the above Eq. (2.139), it is clear that when $n \to \infty$, $e(x) \to 0$.

Furthermore, one very important property of the approximation of a continuous function f on a compact interval $[a, b]$ by orthogonal polynomials is that approximation errors vanish in at least $n + 1$ points of (a, b), where n is the degree of the approximation. We formally state and prove this property as follows.

Theorem 3. *Let $\{\phi_i\}$ be a set of orthogonal polynomials with respect to weight function, w, over compact interval $[a, b]$, where subscript i denotes the degree of the polynomial. Let $\hat{f}$ denote the least squares approximation of a continuous function f using orthogonal polynomials ϕ_i:*

$$\hat{f} = \sum_{i=0}^{n} a_i \phi_i \tag{2.140}$$

Then $f - \hat{f}$ changes sign or vanishes identically at least $n + 1$ times in open interval (a, b).

Proof. The proof of this theorem follows from the most important characteristic of the least squares approximation according to which residual error, $e = f - \hat{f}$, of the least squares solution is orthogonal to the range space spanned by basis functions $\{\phi_i\}$. Now to prove that e must change sign at least $n + 1$ times in (a, b), we first show that e must change sign at least once and then we prove the rest by contradiction arguments.

Note, as e is orthogonal to $\phi_0 = 1$, therefore, $\langle e, 1 \rangle = 0$. Thus, if $e \ne 0$, then it is obvious that e must change sign at least once in (a, b). Now assume that e changes sign fewer than $n + 1$ times and $x_1 < x_2 < \cdots < x_m$ are the points where e changes sign. In each interval (a, x_1), (x_1, x_2), $\cdots$, (x_k, b), e does not change sign but has opposite signs in the neighboring intervals. As a consequence of this, we can define a polynomial function, $P(x) = \prod_{i=1}^{m}(x - x_i)$, of degree m with the following condition:

$$\langle e(x), P(x) \rangle \neq 0 \tag{2.141}$$

However, $P(x)$ being a polynomial of degree $m < n$ can be written as a linear combination of $\phi_0, \phi_1, \cdots, \phi_n$ and is therefore orthogonal to e, i.e., $\langle e(x), P(x) \rangle = 0$. This is a contradiction of Eq. (2.141) and, therefore, e must change sign at least $n + 1$ times in the interval (a, b) $\square$

As a consequence of this theorem, if we use orthogonal polynomials for approximation purposes, then they must interpolate a continuous function exactly ($e = 0$) at $n + 1$ points in the domain of approximation. Note that the key point of the proof of Theorem 3 lies in the fact that residual error e is orthogonal to the range space spanned by basis functions $\{\phi_i\}$.

2.6   Summary

This chapter gives an overview of continuous and discrete orthogonal polynomial basis functions. A significant advantage of using orthogonal functions is that they result in uncoupled linear equations for solution of least squares problems. This feature enables high order approximation without the necessity of inverting matrices, along with many other advantages. Various properties of polynomial approximations are also discussed in this chapter. On several occasions in this text, we will make use of orthogonal polynomial basis functions introduced in this chapter as well as the general methods presented to construct new orthogonal polynomials for novel weight functions. Many of the developments in this text allow users the freedom to select local approximations; when complete freedom exists, the power of orthogonal function approximation should always be considered. Although the discussion here is not exhaustive, this chapter serves to introduce the reader to this very important subject. Interested readers should refer to Ref. [12] for more detailed discussion of approximation theory using orthogonal basis functions.

3

Artificial Neural Networks for Input-Output Approximation*

As for everything else, so for a mathematical theory: beauty can be perceived but not explained.

A. Cayley

3.1 Introduction

Over the past few decades, Artificial Neural Networks (ANNs) have emerged as a powerful set of tools in pattern classification, time series analysis, signal processing, dynamical system modeling and control. The popularity of ANNs can be attributed to the fact that these network models are frequently able to learn behavior when traditional modeling is very difficult to generalize. Typically, a neural network consists of several computational nodes called perceptrons arranged in layers. The number of hidden nodes essentially determines the degrees of freedom of the non-parametric model. A small number of hidden units may not be enough to capture a given system's complex input-output mapping and alternately a large number of hidden units may overfit the data and may not generalize the behavior. It is also natural to ask *"How many hidden layers are required to model the input-output mapping?"* The answer to this question in a general sense is provided by Kolmogorov's theorem [13] (later modified by other researchers [14]), according to which any continuous function from an input subspace to an appropriate output subspace can be approximated by a two-layer neural network with finite number of nodes (model centers).

Kolmogorov's Theorem. *Let $f(\mathbf{x})$ be a continuous function defined on a unit hypercube $\mathbf{I}^n$ ($\mathbf{I} = [0,1]$ and $n \geq 2$), then there exist simple functions***

*©2007 IEEE. Reprinted, with permission, from P. Singla, K. Subbarao, and J. L. Junkins, "Direction-Dependent Learning Approach for Radial Basis Function Networks," *IEEE Transactions on Neural Networks*, vol. 18, no. 1, pp. 203–222.
**Should not be confused with the literal meaning of the word "simple" [13].

ϕ_j and ψ_{ij} such that $f(\mathbf{x})$ can be represented in the following form:

$$f(\mathbf{x}) = \sum_{i=1}^{2n+1} \phi_j \left(\sum_{j=1}^{d} \psi_{ij}(x_j) \right) \tag{3.1}$$

But the main question is how to find these simple functions ϕ_j and ψ_j. In addition to this, the optimal number of hidden units depends upon many factors, like the ability of the chosen basis functions to approximate the given systems behavior, the number of data points, the signal to noise ratio, and the complexity of the learning algorithms. While ANNs are frequently described using network architecture terminology and diagrams, the reality is that any ANN results in a set of parametric interpolation functions representing the input-output behavior. Like any "curve fitting" approach, ANNs must be approached with attention to whether or not the approximation architecture is a good choice for the problem at hand. One must also distinguish between interpolations well supported by neighboring measurements and extrapolation into regions of sparse data or completely outside the region containing measurements. These issues have not been adequately considered in the available literature.

3.1.1   Radial Basis Function Networks

A Radial Basis Function Network (RBFN) is a two-layer neural network that approximates an unknown nonlinear function to represent given input-output data as the weighted sum of a set of radial basis functions

$$f(\mathbf{x}) = \sum_{i=1}^{h} w_i \phi_i(\|\mathbf{x} - \boldsymbol{\mu}_i\|) = \mathbf{w}^T \boldsymbol{\Phi}(\mathbf{x}, \boldsymbol{\mu}_1, \cdots, \boldsymbol{\mu}_h) \tag{3.2}$$

where $\mathbf{x} \in \mathcal{R}^n$ is an input vector, $\boldsymbol{\Phi}$ is a vector of h radial basis functions with $\boldsymbol{\mu}_i \in \mathcal{R}^n$ as the center of the ith radial basis function and $\mathbf{w}$ is a vector of h linear weights or amplitudes. The two layers in an RBFN perform different tasks. The hidden layer with the radial basis function performs a non-linear transformation of the input space into a high dimensional hidden space whereas the outer layer of weights performs the linear regression of the function parameterized by this hidden space to achieve the desired approximation. The linear transformation of Eq. (3.2) of a set of nonlinear basis functions is qualitatively justified by Cover's theorem [15] as follows:

Cover's Theorem. *A complex pattern classification problem or input-output problem cast in a high-dimensional space is more likely to be approximately linearly separable than in a low-dimensional space.*

Cover's theorem provides both a qualitative and a theoretical motivation for using a linear combination of a large number of nonlinear functions to approximate irregular phenomena. According to Cover and Kolmogorov's theorems

[14,15], Multilayered Neural Networks (MLNN) and RBFN can serve as "Universal Approximators," but in actuality, they offer no guarantee on accuracy in practice for a reasonable dimensionality. While MLNN perform a global and distributed approximation at the expense of high parametric dimensionality, RBFNs give a global approximation but with locally dominant basis functions. As an aside, we inject a word of caution before we engage in further developments of the methodology: Both approaches (MLNN and RBFN) potentially suffer from the dual curse of high dimensionality and nonlinearity, and therefore there is a strong motivation to avoid taking excessive comfort in Cover's and Kolmogorov's theorems. These elegant observations provide important qualitative insights from 30,000 feet; however, approximation warfare must often be fought in the trenches. More insight and theoretical developments are required to establish robust algorithms. It is important to view all input-output approximation methods, including all variants of neural networks, in the context of estimation and approximation theory, and avoid panacea-like adoption of universal faith-based cookbooks. Input-output approximation is not a religion—it is in fact just a set of tools to use in multi-dimensional approximation, and as such, we must always be concerned about the distinction between interpolation and extrapolation, and convergence/reliability issues in the presence of sparse and noisy data.

In recent literature [16–19], various choices for radial basis functions are discussed. The Gaussian function is most widely used because, among other reasons, the arguments are points in the space of inputs, and the associated parameters therefore correlate to the local features. These characteristics mean that the network parameters have physical and/or heuristic interpretations. These heuristic local interpretations lead directly to approximations to generate good starting estimates from local measurement data. Applying Gaussian functions to approximate given input-output data can be theoretically supported utilizing the following useful characteristic of the Dirac-Delta function:

$$\delta(f) = \int\limits_{-\infty}^{\infty} \delta_0(x)f(x)dx = f(0) \tag{3.3}$$

In other words, we can think of the above as "$f * \delta \to f$," where "$*$" denotes the convolution operator. Strictly speaking, in the semantics usually adopted for Fourier analysis, $\delta(x)$ is not a function but rather is a distribution [20]. Further, according to the following lemma, such "localized bumps" can be well approximated by Gaussian functions (illustrated in Fig. 3.1.1):

Lemma 1. *Let* $\phi(x) = \frac{1}{\sqrt{2\pi}}e^{-\frac{x^2}{2}}$ *and* $\phi_{(\sigma)}(x) = \frac{1}{\sigma}\phi(\frac{x}{\sigma})$. *If* $C_b(\mathcal{R})$ *denotes the set of continuous, bounded functions over* $\mathcal{R}$, *then*

$$\forall f \in C_b(\mathcal{R}), \lim_{\sigma \to 0} \phi_{(\sigma)} * f(x) = \delta(f) \tag{3.4}$$

Proof. Let us consider $|f(x) - \phi_{(\sigma)} * f(x)|$. Now using the fact that $\int \phi_{(\sigma)} dx = 1$ and the definition of convolution, we have

$$|f(x) - \phi_{(\sigma)} * f(x)| = |\int \phi_{(\sigma)}(y) f(x) dy - \int \phi_{(\sigma)}(y) * f(x - y) dy|$$

$$\leq \int |\phi_{(\sigma)}(y)| |f(x) - f(x - y)| dy$$

Since f is a continuous function, for any given $\epsilon > 0$, there is an $\eta > 0$ such that if $|y| < \eta$ then $|f(x) - f(x - y)| < \epsilon$. This yields the estimate

$$|f(x) - \phi_{(\sigma)} * f(x)| \leq \epsilon \int_{|y| < \eta} |\phi_{(\sigma)}(y)| dy + 2 f_{max} \int_{|y| \geq \eta} |\phi_{(\sigma)}(y)| dy$$

Further, let us compute

$$\int_{|y| \geq \eta} |\phi_{(\sigma)}(y)| dy = \frac{1}{\sigma} \int_{|y| \geq \eta} |\phi(\frac{y}{\sigma})| dy = \int_{|u| \geq \frac{\eta}{\sigma}} |\phi(u)| du$$

Now, this last term tends to 0 as σ tends to 0 since $\eta > 0$. Further, since f is a bounded continuous function, $|f(x) - \phi_{(\sigma)} * f(x)| < \epsilon$ as $\sigma \to 0$ and thus we obtain our desired result as ϵ can be chosen as small as we wish. $\square$

So, theoretically, we can approximate any bounded continuous function with an infinite sum of Gaussian functions, but practically this may lead to a very high dimensioned estimation problem. That said, one can always truncate this infinite sum to some finite number and learn the number of terms required along with other parameters of the Gaussian functions to minimize an appropriate approximation error norm, i.e.,

$$\inf_{\mathbf{p}} \left\{ \|f - \sum_{i=1}^{h} w_i \phi_i(\mathbf{x}, \mathbf{p})\| \right\} \tag{3.5}$$

where $\mathbf{p}$ is a vector of the following free network parameters needed to construct an RBFN:

1. Number of RBFs, h

2. The centers of RBFs, μ_i

3. The spread of RBFs (σ_i in case of Gaussian function)

4. The linear weights between hidden layer and the output layer, w_i

Recently, Narcowich et al. [21] have found Sobolev bounds on approximation error using RBFs as interpolates. More discussion on the approximation characteristics of RBF networks can be found in Refs. [14, 15, 20, 22–24].

In short, the traditional ANNs learning algorithms have serious shortcomings, including:

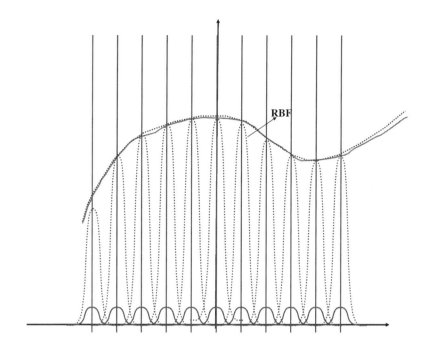

FIGURE 3.1

Illustration of Function Approximation by Localized Bumps and RBF.

1. *Abstraction*: The estimated weights do not have physical significance.

2. *Interpolation versus Extrapolation*: How do we know when a given estimated model is sufficiently well supported such that the network has converged (locally or globally), and has utilized sufficiently dense and accurate measurements in the neighborhood of the desired evaluation point?

3. *Issues Affecting Practical Convergence*: A priori learning versus on-line adaptation? Actually, when the ANN architecture is fixed a priori, then the family of solvable problems is implicitly constrained, which means the *architecture of the network should be learned*, not merely weights adjusted, to ensure efficient and accurate modeling of the particular system behavior.

4. *Uncertainty in Prediction:* There is no exiting methodology that satisfactorily captures the uncertainty in the prediction of the system behavior. How do measurement errors, data distribution, network architecture, and location of the point where the prediction is required couple into errors in the ANN estimates?

The learning methods described in the literature for neural networks seek to minimize the error between network output and observations globally based upon the assumption that all the parameters of the network can be optimized simultaneously. However, the global nature of the distortions can lead to globally optimal network parameters which may minimize the approximation error on the training set but might not be robust when tested on some new data points, or more importantly, used for prediction. The variability of a particular nonlinear system may be nonuniform and nonstationary in space and time; some regions may be highly irregular and others may be smooth and linear. Furthermore, for the case that the neural network is itself nonlinear, the issue of suboptimal convergence to local minima must also be considered. As a consequence, it is improbable that a globally nonlinear input-output mapping parameterization can be guessed a priori that represents such phenomena accurately and efficiently.

In this chapter, we seek to construct an adaptable, intelligent network that is designed such that it seeks to update/learn some or all of the above-mentioned parameters. We will illustrate clearly that it is also vital that the architecture of the network itself be learned in an adaptive way, in order to keep the dimensionality low, and to locate/shape the basis functions to optimize the approximations. To learn various network parameters, different learning algorithms have been suggested in the literature [16–18, 25–30]. Further, adaptation of the architecture of an RBF network, as suggested in Refs. [17, 18, 26, 28, 29, 31], has led to a new class of approximators suitable for multi-resolution modeling applications. While the adaptive nature of these algorithms aids in improving the resolution, it does not necessarily help in the reduction of the number of basis functions required. For all available adaptive RBF networks, the network size can grow indefinitely if high accuracy is sought, due to the fact that the choice of the (fixed) basis function's shape and the the radial basis center distribution over the input space may bear no correlation to the unknown function to be represented. One important root difficulty for most of the methods in the existing literature lies in the fact that the radial basis functions are chosen to be circular (i.e., width of basis function is assumed to be same along all directions), and thus many neighboring circular functions of various sizes must ultimately add and subtract to approximate accurately even moderately non-circular features. Sharp ridges and trenches, as well as highly multi-scale input-output behaviors, are especially challenging to capture accurately and efficiently using circular radial basis functions. In other words, the existing literature provides no consistent means for adaptive reshaping, scaling and rotation of non-circular basis functions to learn from current and past data points. The high degree of redundancy and lack of adaptive reshaping and scaling of RBFs are felt to be serious disadvantages of existing algorithms and provide the motivation for this chapter.

The objectives of this chapter are threefold. First, means for reshaping and rotation of Gaussian functions are introduced to learn the local shape and orientation of the function from the local measurements in a given data set. The orientation of each radial basis function is parameterized through a rotation parameter vector, the magnitude of which for the two- and three-dimensional cases can be shown to be equal to the tangent of the half angle of the principal rotation angle [3]. The principal rotation vector defines the orientation of the principal axes of the quadratic form coefficient function of the Gaussian RBF through parameterization of the principal axes' direction cosine matrix. The shape is captured by solving independently for the principal axis scale factors. We mention that qualitatively, considering a sharp ridge or canyon feature in an input-output map, we can expect the principal axes of the local basis functions to approximately align along and perpendicular to the ridge or canyon. Second, an "intelligent" adaptation scheme is proposed that learns the optimal shape and orientation of the basis functions, along with tuning of the centers and widths to enlarge the size of a single basis function as appropriate to approximate as much of the data as possible. Third, we modify the existing learning algorithms to incorporate the concept of rotation and reshaping of the basis functions to enhance their performance. This objective is achieved by modifying a conventional Modified Resource Allocating Network (MRAN) [17] learning algorithm.

The rest of the chapter is structured as follows: In the next section, the notion of rotation and shape optimization of a Gaussian function in the general case is introduced. Next, a novel learning algorithm is presented to learn the rotation parameters along with the parameters that characterize a regular RBFN. A modification to the MRAN algorithm is introduced to incorporate rotation of the Gaussian basis functions, and finally, the results from various numerical studies are presented to illustrate the efficacy of the algorithms in solving benchmark problems.

3.2   Direction-Dependent Approach

In Ref. [32], we introduce the concept of rotation and reshaping of generally non-circular radial basis functions. Our approach to representing the rotation is motivated through developments in rigid body rotational kinematics [3]. The development is novel because we believe this represents the first application of the rotation ideas to the function approximation problem. We seek the optimal center location, as well as rotation and shape, for the Gaussian basis functions to expand coverage and approximately capture non-circular local behavior, thereby reducing the total number of basis functions required for learning. We mention that this approach can lead to most dramatic improvements when sharp "ridges" or "canyons" exist in the input-output map.

We propose adoption of the following most general n-dimensional Gaussian function

$$\Phi_i(\mathbf{x}, \mu_i, \sigma_i, q_i) = \exp\{-\frac{1}{2}(\mathbf{x} - \boldsymbol{\mu}_i)^T \mathbf{R}_i^{-1}(\mathbf{x} - \boldsymbol{\mu}_i)\} \qquad (3.6)$$

where $\mathbf{R} \in \mathcal{R}^{n \times n}$ is a *fully populated* symmetric positive definite matrix instead of a diagonal one, as in the case of the conventional Gaussian function representation used in various existing learning algorithms. The assumption of a diagonal $\mathbf{R}$ matrix is valid if the variation of output with x_j is uncoupled to x_k, i.e., if different components of the input vector are uncorrelated. In this case, the generalized Gaussian function reduces to the product of n independent Gaussian functions. However, if the parameters are correlated, there are terms in the resulting output that depend on off-diagonal terms of the matrix, $\mathbf{R}$. So it becomes important to learn the off-diagonal terms of the matrix $\mathbf{R}$ for more accurate results (the local basis functions size, shape and orientation can be tailored adaptively to approximate the local behavior).

Now, using spectral decomposition, the matrix $\mathbf{R}^{-1}$ can be written as a similarity transformation product of an orthogonal matrix and a diagonal matrix

$$\mathbf{R}_i^{-1} = \mathbf{C}^T(\boldsymbol{q}_i)\mathbf{S}(\boldsymbol{\sigma}_i)\mathbf{C}(\boldsymbol{q}_i) \qquad (3.7)$$

where $\mathbf{S}$ is a diagonal matrix containing the eigenvalues, σ_{i_k}, of the matrix $\mathbf{R}_i^{-1}$ which dictates the spread of the Gaussian function Φ_i, and $\mathbf{C}(\boldsymbol{q}_i)$ is an $n \times n$ proper orthogonal rotation matrix consisting of eigenvectors of $\mathbf{R}^{-1}$. Now, it is easy to see that contour plots corresponding to a constant value of the generalized Gaussian function, Φ_i, are hyperellipsoids in $\mathbf{x}$-*space*, given by the following equation:

$$(\mathbf{x} - \boldsymbol{\mu}_i)^T \mathbf{R}_i^{-1}(\mathbf{x} - \boldsymbol{\mu}_i) = c^2 \text{ (a constant)} \qquad (3.8)$$

Further, substituting for Eq. (3.7) in Eq. (3.8), we get an equation for another hyperellipsoid in a rotated coordinate system, $\mathbf{y}_i = \mathbf{C}(\mathbf{x} - \boldsymbol{\mu})$:

$$[\mathbf{C}(\boldsymbol{q}_i)(\mathbf{x} - \boldsymbol{\mu}_i)]^T \mathbf{S}(\boldsymbol{\sigma}_i) [\mathbf{C}(\boldsymbol{q}_i)(\mathbf{x} - \boldsymbol{\mu}_i)] = \mathbf{y}_i^T \mathbf{S}(\boldsymbol{\sigma}_i)\mathbf{y}_i = c^2 \text{ (a constant)}$$
$$(3.9)$$

From Eq. (3.9), we conclude that the orthogonal matrix, $\mathbf{C}$, represents the rotation of the orthogonal principle axis of the basis function, Φ_i. Since the eigenvectors of the matrix $\mathbf{R}$ point in the direction of extreme principal axes of the hyperellipsoid, it naturally follows that learning the optimum rotation matrix, $\mathbf{C}$ (whose columns are the eigenvectors of $\mathbf{R}$), is the most helpful in maximal local trend sensing (e.g., to align the principal axis with local features, such as ridges and trenches). Though $\mathbf{C}(\boldsymbol{q}_i)$ is an $n \times n$ squares matrix, we require only $\frac{n(n-1)}{2}$ parameters to describe its most general variation due to the orthogonality constraint ($\mathbf{C}^T\mathbf{C} = \mathbf{I}$). So, in addition to the parameters that characterize a regular RBFN, we now have to adjust the additional parameters characterizing the orthogonal rotation matrix, making a total of

$\frac{(n+2)(n+1)}{2}$ parameters for a minimal parameter description of the most general Gaussian function for an n input single output system. We will find that the apparent increase in the number of parameters is not necessarily a cause for concern because the total number of generalized Gaussian functions required for the representation typically reduces greatly, thereby bringing down the total number of parameters. Also, we will see that the increased accuracy with a reduced number of RBFs provides a powerful heuristic argument for the convergence of this approach. For each RBFN, we require the following parameters:

1. n parameters for the centers of the Gaussian functions, i.e., $\boldsymbol{\mu}$

2. n parameters for the spread (shape) of the Gaussian functions, i.e., $\boldsymbol{\sigma}$

3. $\frac{n(n-1)}{2}$ parameters for a minimal parameterization of the rotation of the principal axis of the Gaussian functions

4. Weight (or amplitude) w_i scaling $\phi_i(.)$'s contribution to the output

Note the existing approaches include only the center locations, spread and weights for circular Gaussian functions. Inclusion of the aforementioned generalized steps 2 and 3 allows the architecture of the network to be learned.

To enforce the positive definiteness and symmetry constraint of matrix $\mathbf{R}$, we propose following three different parameterizations for the covariance matrix, $\mathbf{R}^{-1}$.

1. For the case that we use the spectral decomposition, Eq. (3.7), to represent $\mathbf{R}^{-1}$, we need to parameterize $\mathbf{C}$ as a general proper orthogonal matrix. Hence, we first enforce the orthogonality constraint of the rotation matrix, $\mathbf{C}$, by introducing the following result in matrix theory that is widely used in rotational kinematics, namely, the Cayley Transformation [3]:

   Cayley Transformation. *If $\mathbf{C} \in \mathbb{R}^{n \times n}$ is any proper orthogonal matrix and $\mathbf{Q} \in \mathbb{R}^{n \times n}$ is a skew-symmetric matrix then the following transformations hold:*

   (a) *Forward Transformations*
       i. $\mathbf{C} = (\mathbf{I} - \mathbf{Q})(\mathbf{I} + \mathbf{Q})^{-1}$
       ii. $\mathbf{C} = (\mathbf{I} + \mathbf{Q})^{-1}(\mathbf{I} - \mathbf{Q})$
   (b) *Inverse Transformations*
       i. $\mathbf{Q} = (\mathbf{I} - \mathbf{C})(\mathbf{I} + \mathbf{C})^{-1}$
       ii. $\mathbf{Q} = (\mathbf{I} + \mathbf{C})^{-1}(\mathbf{I} - \mathbf{C})$

   Remarkably, the forward and inverse Cayley transformations are identical. These are among the more beautiful results in linear algebra, yet do not seem as widely utilized as they should be. Since any arbitrary proper orthogonal matrix $\mathbf{C}$ (or skew-symmetric matrix $\mathbf{Q}$) can be substituted into the above-written transformations, the Cayley Transformations can be used to parameterize the entire $\mathbf{O}(n)$ rotational group by skew symmetric matrices. The number of distinct elements in $\mathbf{Q}$ is precisely $\frac{n(n-1)}{2}$, so this is a minimal parameter representation [note there are $\frac{n(n-1)}{2}$ independent constraints implicit in the $2n^2$ orthogonality conditions $\mathbf{C}^T\mathbf{C} = \mathbf{I}, \mathbf{C}\mathbf{C}^T = \mathbf{I}$]. The forward transformation is always well behaved; however, the inverse transformation encounters a difficulty only near the $180°$ rotation where $det\,(\mathbf{I} + \mathbf{C}) \to 0$. Thus $\mathbf{Q}$ is a unique function of $\mathbf{C}$ except at the $180°$ rotation and $\mathbf{C}$ is always a unique function of $\mathbf{Q}$. Thus, as per the Cayley transformation, we can parameterize the orthogonal matrix $\mathbf{C}(\mathbf{q}_i)$ in Eq. (3.7) as

$$\mathbf{C}(\mathbf{q}_i) = (\mathbf{I} + \mathbf{Q}_i)^{-1}(\mathbf{I} - \mathbf{Q}_i) \tag{3.10}$$

where $\mathbf{q}_i$ is a vector of $\frac{n(n-1)}{2}$ distinct elements of a skew symmetric matrix $\mathbf{Q}_i$, i.e., $\mathbf{Q}_i = -\mathbf{Q}_i^T$. Note $\mathbf{q}_i \to 0$ for $\mathbf{C} = \mathbf{I}$ and $-\infty \le \mathbf{q}_i \le \infty$ where $\mathbf{q}_i \to \pm\infty$ corresponds to a $180°$ rotation about any axis. In practice, for the application at hand, we can limit $\mathbf{q}$ to locate all points in a unit hypersphere corresponding to all infinity of $\pm 90°$ rotations. Although by using the Cayley transformation, the orthogonality constraint on the matrix $\mathbf{C}$ can be implicitly guaranteed, one still needs to check for the positive definiteness of $\mathbf{R}$ by requiring $\sigma_i > 0$.

2. Anticipating that various approaches to parameterize $\mathbf{R}$ may have differing computational efficiency in the application of immediate interest, we also introduce the following alternate minimal parameter representation of positive definite matrices. The first is motivated by the definition of a correlation matrix normally encountered in the theory of statistics.

Additive Decomposition. *Let $\mathbf{R} \in \mathcal{R}^{n\times n}$ be a symmetric positive definite matrix, then $\mathbf{R}^{-1}$ is also symmetric and positive definite and can be written as a sum of a diagonal matrix and a symmetric matrix*

$$\mathbf{R}_k^{-1} = \mathbf{\Gamma}_k + \sum_{i=1}^{n}\sum_{j=1}^{n}\mathbf{e}_i\mathbf{e}_j^T q_{k_{ij}} \tag{3.11}$$

where e_i is an $n \times 1$ vector with only the ith element equal to one and the rest of them zeros and $\mathbf{\Gamma}_k$ is a diagonal matrix given by

$$\mathbf{\Gamma}_k = \frac{1}{\sigma_k^2}\mathbf{I} \tag{3.12}$$

subject to the following constraints

$$q_{k_{ij}} = q_{k_{ji}} \tag{3.13}$$

$$\sigma_k > 0 \tag{3.14}$$

$$q_{k_{ii}} > 0 \tag{3.15}$$

$$-1 < \frac{q_{k_{ij}}}{(\sigma_k + q_{k_{ii}})(\sigma_k + q_{k_{jj}})} < 1 \tag{3.16}$$

It is worthwhile to mention that $q_{k_{ij}} \neq 0$ generates the stretching and rotation of the Gaussian function. If $q_{k_{ij}} = 0$ then we obviously obtain the circular Gaussian function. Even though the learning of the matrix, $\mathbf{R}$, is greatly simplified by this parameterization, one needs to impose the constraints defined in Eqs. (3.13)−(3.16) during the parameter learning process.

3. To explicitly enforce the positive definiteness and symmetry of the co-variance matrix, $\mathbf{R}^{-1}$, one could alternatively use the Cholesky decomposition [3].

Cholesky Decomposition. *Let $\mathbf{R} \in \mathcal{R}^{n \times n}$ be a symmetric positive definite matrix, then $\mathbf{R}^{-1}$ is also symmetric and positive definite and can be factored into a lower triangular matrix times its transpose such that*

$$\mathbf{R}^{-1} = \mathbf{L}\mathbf{L}^T \tag{3.17}$$

where $\mathbf{L}$ is an lower triangular matrix given by the following expression:

$$\mathbf{L} = \begin{bmatrix} l_{11} & 0 & 0 & \cdots & 0 \\ l_{21} & l_{22} & 0 & 0 & 0 \\ \vdots & \vdots & \vdots & \ddots & \vdots \\ l_{n1} & l_{n2} & l_{n3} & \cdots & l_{nn} \end{bmatrix}$$

Notes: The Cholesky upper triangular matrix, $\mathbf{L}^T$, is also known as the matrix squares root of positive definite matrix, $\mathbf{R}^{-1}$. There are $n + \frac{n(n-1)}{2}$ distinct elements in $\mathbf{L}$, so we note Eq. (3.17) is another elegant minimal parameter representation of $\mathbf{R}^{-1}$.

Based on numerical studies we have conducted to date, the Cholesky decomposition parameterization of the matrix $\mathbf{R}^{-1}$ is computationally more attractive than the other two parameterizations in numerical optimization because the symmetry and positive definiteness properties of $\mathbf{R}^{-1}$ are explicitly enforced in this case to get rid of any kind of constraints. However, to our knowledge, the use of any of the three above parameterizations for aiding parameter updating in radial basis function network approximation applications is an innovation introduced in Ref. [32]. We have experimented with all three approaches and while studies to date favor the Cholesky decomposition (mainly because of programming convenience), the other two representations will likely be found advantageous in some future applications. Regarding the generalization of RBFN to include the use of rotations, preliminary studies indicate a very significant reduction in the number of basis functions required to accurately model unknown functional behavior of the actual input-output data. In the subsequent sections, we report a novel learning algorithm and a modified version of the MRAN algorithm to adapt recursively this extended set of parameters. We also report the results of applications to five benchmark problems and provide comparisons with existing algorithms.

3.3   Directed Connectivity Graph

A common main feature of the proposed learning algorithms is a judicious starting choice for the location of the RBF. We introduce here a Directed Connectivity Graph (DCG) approach; this allows a priori adaptive sizing of the network for off-line learning, location and shaping of the dominant local RBFs, and zeroth order network pruning. Because the Gaussian RBFN is a nonlinear representation, we know that finding the RBFN parameter values that locate the global minimum approximation error is challenging and generally iterative. For this reason, means to initiate learning with a good approximation is very important. It also seems qualitatively important to begin with a low dimensional model with a small number of RBFs that capture macroscopic dominant trends and then adaptively inject additional granular RBFs to capture increasingly localized features in the input-output map. Direction-dependent scaling and rotation of basis functions are initialized for maximal local trend sensing with minimal parameter representations and adaptation of the network parameters is implemented to allow for on-line tuning.

The first step toward obtaining a zeroth order off-line model is the judicious selection of a set of basis functions and their center locations, followed by proper initialization of the shape and orientation parameters. This exercise is the focus of this section.

To choose the locations for the RBF centers, we make use of following Lemma that essentially states that *"the center of a Gaussian function is an extremum point."*

Lemma 2. *Let* $\Phi(\mathbf{x}) : \mathcal{R}^n \to \mathcal{R}$ *represent a Gaussian function, i.e.,* $\Phi(\mathbf{x}) = \exp\left(-\frac{1}{2}(\mathbf{x} - \boldsymbol{\mu})^T \mathbf{R}^{-1}(\mathbf{x} - \boldsymbol{\mu})\right)$ *then* $\mathbf{x} = \boldsymbol{\mu}$ *is the only extremum point of* $\Phi(\mathbf{x})$, *i.e.,* $\frac{d\Phi}{d\mathbf{x}}\big|_{\mathbf{x}=\boldsymbol{\mu}} = 0$. *Further,* $\mathbf{x} = \boldsymbol{\mu}$ *is the global maximum of* Φ.

Proof. This lemma is pretty obvious, but formally we see the gradient of Φ is

$$\frac{d\Phi}{d\mathbf{x}} = \exp\left(-\frac{1}{2}(\mathbf{x} - \boldsymbol{\mu})^T \mathbf{R}^{-1}(\mathbf{x} - \boldsymbol{\mu})\right)\mathbf{R}^{-1}(\mathbf{x} - \boldsymbol{\mu}) \qquad (3.18)$$

Now, since $\mathbf{R}^{-1}$ is a positive definite symmetric covariance matrix, from equation (3.18), it is clear that $\frac{d\Phi}{d\mathbf{x}} = 0$ iff $\mathbf{x} = \boldsymbol{\mu}$. Further, it is easy to check that

$$\frac{d \log \Phi}{d\mathbf{x}} = -\mathbf{R}^{-1}(\mathbf{x} - \boldsymbol{\mu}) \tag{3.19}$$

$$\nabla^2 \log \Phi(\mathbf{x}) = -\mathbf{R}^{-1} \tag{3.20}$$

Since $\frac{d \log \Phi}{d\mathbf{x}}|_{\mathbf{x}=\boldsymbol{\mu}} = 0$ and $\nabla^2 \log \Phi(\mathbf{x}) < 0$ we conclude that $\mathbf{x} = \boldsymbol{\mu}$ is the only maximum point of $\log \Phi$. Since log is a monotonically increasing function of $(\mathbf{x} - \boldsymbol{\mu})$, so the center of the Gaussian function, $\boldsymbol{\mu}$, is also a global maximum point of the Gaussian function. $\qquad \square$

Thus, from the above-mentioned lemma, all the interior extremum points of the given surface data should heuristically be the first choice for location of Gaussian functions with the $\mathbf{R}$ matrix determined to first order by the covariance of the data confined in a judicious local mask around each particular extremum point. Therefore, the first step of the learning algorithm for an RBFN should be *to approximately locate the extremum points of a given input-output map*. It should be noted that since the functional expression for the input-output map is unknown, to find approximate extremum points from discrete surface data, we need to check the necessary condition that the first derivative of the unknown input-output map should be zero at each and every data point. We mention that the process of checking this condition at every data point is very tedious and computationally expensive, and for discrete data, requires local approximations to estimate the partial derivatives. Note it is not difficult to test for relative extrema of adjacent function values by direct comparison. Lemma 3 provides an efficient algorithm to approximate the extremum points of the given input-output data.

Lemma 3. *Let $f : \mathcal{X} \to \mathcal{R}$ be a continuous function, where $\mathcal{X}$ is a paracompact space with $\mathcal{U} = \{U_\alpha\}_{\alpha \in \mathcal{A}}$ as an open covering, i.e., $\mathcal{X} \subset \cup_{\alpha \in \mathcal{A}} U_\alpha$. If $\mathcal{S}$ denotes the set of all extremum points of f then there exists a refinement, $\mathcal{V} = \{V_\beta\}_{\beta \in \mathcal{B}}$, of the open covering $\mathcal{U}$, such that $\mathcal{S} \subseteq \mathcal{W}$, where $\mathcal{W}$ is the set of the relative maxima and minima of f in open sets V_α.*

Proof. The proof of this lemma follows from the fact that the input space $\mathcal{X}$ is a paracompact space because it allows us to refine any open cover $\mathcal{U} = \{U_\alpha\}_{\alpha \in \mathcal{A}}$ of $\mathcal{X}$. Let $\mathcal{U} = \{U_\alpha\}_{\alpha \in \mathcal{A}}$ be the open cover of the input space $\mathcal{X}$. Further, assume that x_{max_α} and x_{min_α} define the maximum and minimum values of f in each open set U_α, respectively, and $\mathcal{W}$ is the set of all such points, i.e., $card(\mathcal{W}) = 2card(\mathcal{A})$. Now, we know that the set of all local maximum and minimum points of any function is the same as the set $\mathcal{S}$ of extremum points of that function. Further, without loss of generality we can assume that the set, $\mathcal{W}$, of all local maxima and minima of the function in each open set, U_α, is a subset of $\mathcal{S}$ because if it is not, then we can refine the open cover $\mathcal{U}$ further until this is true. $\qquad \square$

According to Lemma 3, for mesh sizes less than a particular value, the set $\mathcal{S}$, of the extremum points of the unknown input-output map f, should be in a subset of the set $\mathcal{W}$, consisting of the relative maxima and minima of the data points in each grid element. Now, the set $\mathcal{S}$ can be extracted from set $\mathcal{W}$ by checking the necessary condition that the first derivative of f should be zero at extremum points. This way one need only approximate the first derivative of the unknown map at $2M$ points, where M is the total number of elements in which data have been divided. Alternatively, we can simply adopt any relative extreme data points in the elements as the local extrema. It should be noted that M is generally much smaller than the total number of data points available to approximate the unknown input-output map. We mention that these approximate locations are only good starting estimates for an optimization process to be described below, and therefore we can resist the temptation to rigorously locate extremum points to high precision.

Further, to systematically introduce new centers from the set $\mathcal{S}$, we construct directed graphs $\mathcal{M}$ and $\mathcal{N}$ of all the relative maxima sorted in descending order and all the relative minima sorted in ascending order, respectively. We then choose the points in $\mathcal{M}$ and $\mathcal{N}$ sequentially as candidates for Gaussian function centers with the extreme function value as the corresponding starting weight of the Gaussian functions. The centers at the points in $\mathcal{M}$ and $\mathcal{N}$ are introduced recursively until some convergence criteria is satisfied. The initial value of each local covariance matrix $\mathbf{R}$ is approximated from the statistical covariance of the data in a local mask around the chosen center. Now, using all the input data, we adapt the parameters of the chosen Gaussian functions and, upon convergence, check the error residuals for the estimation error. If the error residuals norm is larger than a predefined bound, we choose the next set of points in the directed graphs $\mathcal{M}$ and $\mathcal{N}$ as center locations for additional Gaussian RBFs and repeat the whole process. The network only grows in dimensionality when error residuals cannot be made sufficiently small, and thus the increased dimensionality grows only incrementally with the introduction of a judiciously shaped and located basis function. Finally, the set of RBFs can be enlarged beyond those associated with starting centers and shapes from the $\mathcal{M}$ and $\mathcal{N}$ sets, driven by the location of maxima and minima of residual errors in the input-output approximation. The refinement process can be continued indefinitely until the largest residual error is less than a prescribed bound. The initial location and shape parameters are simply the starting estimates for the learning algorithm; we show below that the combination of introducing basis functions sequentially and estimating their shape and location from local data is highly effective with regard to obtaining high accuracy with a small number of RBFs and network parameters.

3.3.1   Estimation Algorithm

The heart of any learning algorithm for RBFN is an estimation algorithm to adapt initially defined network parameters so that approximation errors are reduced to smaller than some specified tolerance. Broadly speaking, none of the nonlinear optimization algorithms available guarantee that the global optimum will be achieved. Estimation algorithms based on the least squares criteria are the most widely used methods for estimation of the constant parameter vector from a set of redundant observations. According to the least squares criteria, the optimum parameter value is obtained by minimizing the sum of squares of the vertical offsets (Residuals) between the observed and computed approximations. In general, for nonlinear problems, successive corrections are made based upon local Taylor series approximations. Further, any estimation algorithm generally falls into the category of either a batch estimator or a sequential estimator, depending upon the way in which observation data is processed. A batch estimator usually processes a large batch of data taken from a fixed span of the independent variable (usually time) to estimate the optimum parameter vector, while a sequential estimator is based upon a recursive algorithm, which updates the parameter vector in a recursive manner after receipt of each observation. Due to their recursive nature, sequential estimators are preferred for real time estimation problems; however, batch estimators are frequently preferable for off-line learning.

To adapt the various parameters of the RBFN as defined in the previous section, we use an extended Kalman filter [33] for on-line learning, while the Levenberg-Marquardt [34,35] batch least squares algorithm is used for off-line learning. Kalman filtering is a modern (since 1960) development in the field of estimation [2,6] though it has its roots as far back as Gauss' work in the 1800s. In the present study, the algebraic version of the Kalman filter is used since our model does not involve differential equations. On the other hand, the Levenberg-Marquardt estimator, being the combination of the *method of steepest descent* and *the Gauss-Newton method of differential correction*, is a powerful batch estimator tool in the field of nonlinear least squares [2]. We mention that both the algorithms are very attractive for the problem at hand, and details of both the algorithms can be found in Ref. [2]. Further, for some problems, the Kalman filter is attractive as a means to update the off-line a priori learned network parameters in real time whenever new measurements are available. The implementation equations for the extended Kalman filter or *Kalman-Schmidt filter* are given in Table 3.1.

Since the covariance update is based upon an assumption of linearity, it is typically useful to impose a lower bound on the eigenvalues of P_k^+ to keep the Kalman filter from becoming too optimistic and rejecting new measurements. To learn the different parameters of the RBFN using any estimation algorithm, the sensitivity (Jacobian) matrix $\mathbf{H}$ needs to be computed. The various partial derivatives required to synthesize the sensitivity matrix are outlined in subsequent subsections for all three parameterizations of $\mathbf{R}^{-1}$ described in Section 3.2.

TABLE 3.1
Kalman-Schmidt Filter

Measurement Model	$$\tilde{\mathbf{y}} = \mathbf{h}(\mathbf{x}_k) + \boldsymbol{\nu}_k$$ with $$\mathbf{E}(\boldsymbol{\nu}_k) = 0$$ $$\mathbf{E}(\boldsymbol{\nu}_l \boldsymbol{\nu}_k^T) = \mathbf{R}_k \delta(l - k)$$
Update	$$\mathbf{K}_k = \mathbf{P}_k^- \mathbf{H}_k^T (\mathbf{H}_k \mathbf{P}_k^- \mathbf{H}_k^T + \mathbf{R}_k)^{-1}$$ $$\hat{\mathbf{x}}_k^+ = \hat{\mathbf{x}}_k^- + \mathbf{K}_k(\tilde{\mathbf{y}} - h(\mathbf{x}_k^-))$$ $$\mathbf{P}_k^+ = (\mathbf{I} - \mathbf{K}_k \mathbf{H}) \mathbf{P}_k^-$$ where $$\mathbf{H}_k = \frac{\partial \mathbf{h}(\mathbf{x}_k)}{\partial \mathbf{x}}\Big|_{\mathbf{x} = \hat{\mathbf{x}}_k^-}$$

3.3.2 Spectral Decomposition of the Covariance Matrix

Using Eq. (3.7), we represent $\mathbf{R}^{-1}$ using the spectral decomposition and adopt the typical RBF as the exponential of Eq. (3.6). In this case, the sensitivity matrix, $\mathbf{H}$, can be defined as

$$\mathbf{H} = \frac{\partial f(\mathbf{x}, \mu, \sigma, \mathbf{q})}{\partial \Theta} \tag{3.21}$$

where $f(\mathbf{x}, \boldsymbol{\mu}, \boldsymbol{\sigma}, \mathbf{q}) = \sum_{i=1}^{N} w_i \Phi_i(\boldsymbol{\mu}_i, \boldsymbol{\sigma}_i, \mathbf{q}_i)$ and Θ is an $N \times \frac{(n+1)(n+2)}{2}$ vector given by

$$\Theta = \left\{ w_1 \ \boldsymbol{\mu}_1 \ \boldsymbol{\sigma}_1 \ \mathbf{q}_1 \ \cdots \ w_N \ \boldsymbol{\mu}_N \ \boldsymbol{\sigma}_N \ \mathbf{q}_N \right\} \tag{3.22}$$

Here, $\mathbf{q}$ is a $\frac{n(n-1)}{2}$ vector used to parameterize the rank deficient skew-symmetric matrix $\mathbf{Q}$ in Eq. (3.10).

$$Q_{ij} = 0, \ i = j \tag{3.23}$$
$$= q_k \ i < j$$

where $k = \|i - j\|$ if $i = 1$ and $k = \|i - j\| + \|i - 1 - n\|$ for $i > 1$. Note that the lower triangular part of $\mathbf{Q}$ can be formed using the skew-symmetry property of $\mathbf{Q}$. The partial derivatives required for the computation of the sensitivity matrix, $\mathbf{H}$, are obtained using Eqs. (3.6), (3.7) and (3.10), as follows:

$$\frac{\partial f}{\partial w_k} = \phi_k \tag{3.24}$$

$$\frac{\partial f}{\partial \boldsymbol{\mu}_k} = \left[w_k \phi_k \mathbf{R}_k^{-1}(\mathbf{x} - \boldsymbol{\mu}_k)\right]^T \tag{3.25}$$

$$\frac{\partial f}{\partial \boldsymbol{\sigma}_{k_i}} = w_k \phi_k \frac{\mathbf{y}_i^2}{\sigma_{k_i}^3}, \mathbf{y}_i = \mathbf{C}_k(\mathbf{x} - \boldsymbol{\mu}_k), \ i = 1 \dots n \tag{3.26}$$

$$\frac{\partial f}{\partial \mathbf{q}_{k_l}} = -\frac{w_k}{2}\phi_k \left[(\mathbf{x} - \boldsymbol{\mu}_k)^T \frac{\partial \mathbf{C}_k^T}{\partial \mathbf{q}_{k_l}} \mathbf{S}_k \mathbf{C}_k(\mathbf{x} - \boldsymbol{\mu}_k)\right.$$
$$\left. + (\mathbf{x} - \boldsymbol{\mu}_k)^T \mathbf{C}_k^T \mathbf{S}_k \frac{\partial \mathbf{C}_k}{\partial \mathbf{q}_{k_l}}(\mathbf{x} - \boldsymbol{\mu}_k)\right], l = 1 \dots n(n-1)/2 \tag{3.27}$$

Further, the partial $\frac{\partial \mathbf{C}_k^T}{\partial \mathbf{q}_{k_l}}$ in Eq. (3.27) can be computed by substituting for $\mathbf{C}$ from Eq. (3.10):

$$\frac{\partial \mathbf{C}_k}{\partial \mathbf{q}_{k_l}} = \frac{\partial}{\partial \mathbf{q}_{k_l}}(\mathbf{I} + \mathbf{Q}_k)^{-1}(\mathbf{I} - \mathbf{Q}_k) + (\mathbf{I} + \mathbf{Q}_k)^{-1}\frac{\partial}{\partial \mathbf{q}_{k_l}}(\mathbf{I} - \mathbf{Q}_k) \tag{3.28}$$

Making use of the fact that $(\mathbf{I} + \mathbf{Q})^{-1}(\mathbf{I} + \mathbf{Q}) = \mathbf{I}$, we get

$$\frac{\partial}{\partial \mathbf{q}_{k_l}}(\mathbf{I} + \mathbf{Q}_k)^{-1} = -(\mathbf{I} + \mathbf{Q}_k)^{-1}\frac{\partial \mathbf{Q}_k}{\partial \mathbf{q}_{k_l}}(\mathbf{I} + \mathbf{Q}_k)^{-1} \tag{3.29}$$

Substitution of Eq. (3.29) in Eq. (3.28) gives

$$\frac{\partial \mathbf{C}_k}{\partial \mathbf{q}_{k_l}} = -(\mathbf{I} + \mathbf{Q}_k)^{-1}\frac{\partial \mathbf{Q}_k}{\partial \mathbf{q}_{k_l}}(\mathbf{I} + \mathbf{Q}_k)^{-1}(\mathbf{I} - \mathbf{Q}_k) - (\mathbf{I} + \mathbf{Q}_k)^{-1}\frac{\partial \mathbf{Q}_k}{\partial \mathbf{q}_{k_l}} \tag{3.30}$$

Now, Eqs. (3.24)–(3.27) constitute the sensitivity matrix $\mathbf{H}$ for the extended Kalman filter. We mention that although Eq. (3.7) provides a minimal parameterization of the matrix $\mathbf{R}$, we need to make sure that the scaling parameters denoted by σ_i are always greater than zero. So in case of any violation of this constraint, we need to invoke the parameter projection method to project inadmissible parameters onto the boundary of the set they belong to, thereby ensuring that the matrix $\mathbf{R}$ remains symmetric and positive definite at all times. Further, based on our experience with this parameterization, it is highly nonlinear in nature and sometimes causes unreliable convergence of the estimation algorithm. We found that this difficulty is alleviated by considering the two alternate representations discussed earlier. We summarize the sensitivity matrices for these alternate parameterizations in the next subsections.

3.3.3 Additive Decomposition of the Covariance Matrix

Using the additive decomposition for the $\mathbf{R}_i$ matrix in Eq. (3.6) the different partial derivatives required for synthesizing the sensitivity matrix $\mathbf{H}$ can be computed. We define the following parameter vector $\boldsymbol{\Theta}$

$$\boldsymbol{\Theta} = \left\{ w_1\ \boldsymbol{\mu}_1\ \sigma_1\ \mathbf{q}_1\ \cdots\ w_N\ \boldsymbol{\mu}_N\ \sigma_N\ \mathbf{q}_N \right\} \tag{3.31}$$

The required partials with respect to the elements of $\boldsymbol{\Theta}$ are then given as follows

$$\frac{\partial f}{\partial w_k} = \phi_k \tag{3.32}$$

$$\frac{\partial f}{\partial \boldsymbol{\mu}_k} = \left[w_k \phi_k \mathbf{P}_k^{-1}(\mathbf{x} - \boldsymbol{\mu}_k) \right]^T \tag{3.33}$$

$$\frac{\partial f}{\partial \sigma_{k_i}} = w_k \phi_k \frac{(\mathbf{x}_i - \boldsymbol{\mu}_{k_i})^2}{\sigma_{k_i}^3}, i = 1 \ldots n \tag{3.34}$$

$$\frac{\partial f}{\partial \mathbf{q}_{k_l}} = -w_k \phi_k (\mathbf{x}_i - \boldsymbol{\mu}_{k_i})^T (\mathbf{x}_j - \boldsymbol{\mu}_{k_j}),\ l = 1 \ldots n(n+1)\backslash 2, i, j = 1 \ldots n \tag{3.35}$$

Thus, Eqs. (3.32)−(3.35) constitute the sensitivity matrix $\mathbf{H}$. It should be mentioned that even though the synthesis of the sensitivity matrix is greatly simplified, one needs to check the constraints defined in Eqs. (3.13)−(3.16) at every update. In case these constraints are violated, we once again invoke the parameter projection method to project the parameters normal to the constraint surface to the nearest point on the set they belong to, thereby ensuring that the covariance matrix remains symmetric and positive definite at all times.

3.3.4 Cholesky Decomposition of the Covariance Matrix

As in the previous two cases, once again the sensitivity matrix, $\mathbf{H}$, can be computed by defining the parameter vector, $\boldsymbol{\Theta}$, as

$$\boldsymbol{\Theta} = \left\{ w_1\ \boldsymbol{\mu}_1\ \mathbf{l}_1\ \cdots\ w_n\ \boldsymbol{\mu}_n\ \mathbf{l}_n \right\} \tag{3.36}$$

where $\mathbf{l}_i$ is the vector of elements parameterizing the lower triangular matrix, $\mathbf{L}$.

Carrying out the algebra, the required partials can be computed as

$$\frac{\partial f}{\partial w_k} = \phi_k \tag{3.37}$$

$$\frac{\partial f}{\partial \boldsymbol{\mu}_k} = \left[w_k \phi_k \mathbf{R}_k^{-1}(\mathbf{x} - \boldsymbol{\mu}_k) \right]^T \tag{3.38}$$

$$\frac{\partial f}{\partial \mathbf{l}_{k_l}} = -\frac{w_k}{2} \phi_k \left[(\mathbf{x} - \boldsymbol{\mu}_k)^T \left(\frac{\partial \mathbf{L}_k}{\partial \mathbf{l}_{k_l}} \mathbf{L}_k^T + \mathbf{L}_k \frac{\partial \mathbf{L}_k^T}{\partial \mathbf{l}_{k_l}} \right) (\mathbf{x} - \boldsymbol{\mu}_k) \right],$$
$$l = 1 \ldots n(n-1)\backslash 2 \tag{3.39}$$

Further, $\mathbf{L}_k$ can be written as

$$\mathbf{L}_k = \sum_{i=1}^{n} \sum_{j=i}^{n} e_i e_j L_{k_{ij}} \tag{3.40}$$

Therefore, $\frac{\partial \mathbf{L}_k}{\partial l_{k_l}}$ can be computed as

$$\frac{\partial \mathbf{L}_k}{\partial l_{k_l}} = \sum_{i=1}^{n} \sum_{j=i}^{n} e_i e_j \tag{3.41}$$

Thus, Eqs. (3.37)−(3.39) constitute the sensitivity matrix, $\mathbf{H}$. It should be mentioned that unlike the Cayley transformation and the additive decomposition, Cholesky decomposition guarantees the symmetry and positive definiteness of the matrix, $\mathbf{R}^{-1}$, without additional constraints and so is more attractive for learning the matrix, $\mathbf{R}^{-1}$.

It should be noted that although these partial derivatives are computed to synthesize the sensitivity matrix for the extended Kalman filter, they are required in any case, even if a different parameter estimation algorithm is used (i.e., the computation of these sensitivity partials is inevitable).

The steps for implementing the Directed Connectivity Graph Learning Algorithm are summarized as follows:

Step 1 Find the interior extremum points, i.e., approximate locations of the global maximum and minimum of the given input-output data.

Step 2 Grid the given input space, $\mathcal{X} \in \mathcal{R}^n$, using hypercubes of length l.

Step 3 Find the relative maximum and minimum of given input-output data on the grid points in the region covered by each hypercube.

Step 4 Make a directed graph of all maximum and minimum points sorted in descending and ascending order, respectively. Denote the directed graph of maximum points and minimum points by $\mathcal{M}$ and $\mathcal{N}$, respectively.

Step 5 Choose the first point from graphs $\mathcal{M}$ and $\mathcal{N}$, denoted by $\mathbf{x}_M$ and $\mathbf{x}_N$, respectively, as candidates for Gaussian center and respective measured output function values as the initial weight estimate of those Gaussian functions because at the center the Gaussian function response is 1.

Step 6 Approximate the initial covariance matrix estimate, $\mathbf{R}$, directly from the computed statistical covariance matrix using the observations in a specified size local mask around points $\mathbf{x}_M$ and $\mathbf{x}_N$.

Step 7 Parameterize the covariance matrix, $\mathbf{R}$, using one of the three parameterizations defined in Section 3.2.

Step 8 Use the extended Kalman filter or the Levenberg-Marquardt algorithm to refine the parameters of the network using the given input-output data.

Step 9 On each iteration, use parameter projection to enforce parametric constraints, if any, depending upon the covariance matrix decomposition.

Step 10 Check the estimation error residuals. If they do not satisfy the prescribed accuracy tolerance, then choose the next point in the directed graphs $\mathcal{M}$ and $\mathcal{N}$ as the Gaussian center and restart at step 5.

Step 11 If all the points in sets $\mathcal{M}$ and $\mathcal{N}$ have been used to refine the approximation, and the maximum residual error still exceeds the tolerance, then locate a new center at the location of the maximum residual error, with residual error covariance used as a starting estimate for $\mathbf{R}_k$.

Step 12 Repeat step 11 until the prescribed error tolerance is satisfied.

The grid generation in step 2 is computationally costly unless careful attention is paid to efficiency. To grid the input space $\mathcal{X} \in \mathcal{R}^n$ in a computationally efficient way, we designate a unique cell number to each input point in N^{th} decimal system, depending upon its coordinates in $\mathcal{R}^n$. Here, $N = \max\{N_1, N_2, \cdots, N_n\}$ and N_i denotes the number of cells required along the ith direction. The pseudo-code for the grid generation is given below.

Psuedo-Code for Grid Generation

```
for  ct=1:n
       xlower(ct)=min(indata(:,ct))
       xupper(ct)=max(indata(:,ct))
end
deltax=(xupper-xlower)/N
for  ct=1:Npoints
       cellnum(ct)=ceil((indata(ct,:)-xlower)/deltax)
       cellindex(ct)=getindex(cellnum(ct))
end
```

The relative maxima and minima in each cell are calculated by using all the data points with the same cell number. Though this divide and conquer process of finding the centers and evaluating the local covariance followed by the function evaluation with adaptation and learning may seem to be computationally extensive, it helps in dramatically reducing the total number of Gaussian functions, and therefore helps keep the *"curse of dimensionality"* in check. Further, the rotation parameters and shape optimization of the Gaussian functions enable us to approximate the local function behavior with greatly improved accuracy. Finally, we offer the qualitative observation: We have found that using this process to initialize the learning process greatly improves the approximation process since we begin with a representation that already approximates the macroscopic input-output map. Since we use the Kalman filter to refine the parameters of the RBF network, the selection of starting estimates for the centers can be made off-line with some training data, and the same algorithm can be invoked on-line as new measurements are processed to adapt the parameters from the off-line (a priori) network. Obviously, we can choose to constrain any subset of the network parameters, if necessary, to implicitly obtain a suboptimal approximation but with reduced dimensionality. Any new Gaussian centers can be added to the existing network. These can be introduced locally and adaptively based upon the statistical information of the approximation errors. Additional localization and reduction in the computational burden can be achieved by exploiting the local dominance near a given point by adjusting only a small subset of locally dominant RBFN parameters. Thus there are several avenues to continue convergence progress when dimensionality-induced computational challenges arise.

3.4 Modified Minimal Resource Allocating Algorithm (MMRAN)

In this section, we illustrate how the rotation and reshaping parameters can be incorporated into existing RBF learning algorithms by modifying the popular Minimal Resource Allocating Network (MRAN). To show the effectiveness of this modification, we also include the rotation parameters as adaptable parameters, while keeping the same center selection and pruning strategy as in the conventional MRAN. For the sake of completeness, we give a brief introduction to MRAN. The reader should refer to Ref. [17] for more details. (Note that MRAN is generally accepted as a significant improvement of the Resource Allocating Network (RAN) of Platt [29].) MRAN adopts the basic idea of adaptively "growing" the number of radial basis functions where needed to null local errors, and also includes a "pruning strategy" to eliminate little-needed radial basis functions (those with weights smaller than some tolerance), with the overall goal of finding a minimal RBF network. RAN allocates new units as well as adjusts the network parameters to reflect the complexity of function being approximated. The problem of allocating RBF basis functions sequentially was stated as follows in Ref. [26]: "*Given the prior approximation f^{n-1} and the new observation (x_n, y_n), how do we combine these two information sets to obtain the posterior approximation f^n?*" The optimal approximation for f^n is to add an impulse function at $\mathbf{x}_n$ to f^{n-1} which compensates for the difference in the estimated response and the actual response.

$$f^n(\mathbf{x}) = f^{n-1}(\mathbf{x}) + \delta_n(y_n - f^{n-1}(\mathbf{x}_n)) \tag{3.42}$$

This approach will ensure that the existing features of a prior network are maintained and error for the new added unit is zero. But such a solution lacks smoothness of the underlying function. We might anticipate that this approach is also prone to error when the new measurement contains measurement errors. Therefore, we choose to use Gaussian functions centered at $\mathbf{x}_n$ instead of an impulse function to get a smooth approximation.

$$\phi_n(\mathbf{x}) = \exp(-\frac{1}{\sigma_n^2}\|\mathbf{x} - \mathbf{x}_n\|^2) \tag{3.43}$$

If we let the number of hidden units required to approximate f^{n-1} be h, then we can write

$$
\begin{aligned}
f^n(\mathbf{x}) &= \sum_{i=1}^{h} w_i\phi_i(\mathbf{x}) + (y_n - f^{n-1}(\mathbf{x}_n))\phi_n(\mathbf{x}) \\
&= \sum_{i=1}^{h+1} w_i\phi_i(\mathbf{x})
\end{aligned}
\tag{3.44}
$$

Therefore the parameters associated with the new hidden unit are given as follows:

$$w_{h+1} = y_n - f^{n-1}(\mathbf{x}_n) \tag{3.45}$$
$$\boldsymbol{\mu}_{h+1} = \mathbf{x}_n \tag{3.46}$$
$$\sigma_{h+1} = \sigma_n \tag{3.47}$$

Heuristically, the estimated width of the new Gaussian function, σ_n, is chosen in MRAN to be proportional to the shortest distance between $\mathbf{x}_n$ and the existing centers, i.e.,

$$\sigma_n = \kappa\|\mathbf{x}_n - \boldsymbol{\mu}_{nearest}\| \tag{3.48}$$

κ should be chosen judiciously to account for the amount of overlap between different Gaussian functions.

Even though the above approach is better than simply adopting a fixed architecture, the main difficulty is that we may go on adding new hidden units that contribute little to the final estimate. Therefore, a new hidden unit is actually added to the existing network only if it satisfies the following criteria [29]:

$$\|\mathbf{x}_i - \boldsymbol{\mu}_{nearest}\| > \epsilon \tag{3.49}$$
$$\|e_i\| = \|y_i - f(\mathbf{x}_i)\| > e_{min} \tag{3.50}$$
$$e_i^{rms} = \sqrt{\sum_{j=i-(N_w-1)}^{i} \frac{\|e_j\|^2}{N_w}} > e_{r_{min}} \tag{3.51}$$

Eq. (3.49) ensures that a new RBF node is added if it is sufficiently far from all the existing nodes. If the inequality of Eq. (3.50) is satisfied, then the the approximation error using existing nodes meet the error specification and no new node is added. Eq. (3.51) takes care of noise in the observations by checking the sum squared error for past N_w observations. ϵ, e_{min} and $e_{r_{min}}$ are different thresholds which should be chosen appropriately to achieve desired accuracy.

If the above-mentioned criteria are not met, then the following network parameters are updated using the gradient descent approach or extended Kalman filter as suggested by Sundararajan [17].

$$\Theta = \left\{ w_1 \; \boldsymbol{\mu}_1^T \; \sigma_1 \; \cdots \; w_h \; \boldsymbol{\mu}_h^T \; \sigma_h \right\} \tag{3.52}$$

The advantages of MRAN over other learning algorithms can be summarized as follows:

- It is inherently sequential in nature and therefore can be used recursively in real-time to update the estimated model.

- The network architecture itself is adapted in contrast to simply adjusting weights in a fixed architecture network. The ability of the network to capture the input-output behavior typically improves as more measurements are available.

The adaptive architecture feature and the inherent recursive structure of the learning algorithm make this approach ideal for multi-resolution modeling [26, 31, 36]. While the methodology is very effective in some cases, it still suffers from the drawback of potential explosion in the number of basis functions required to approximate the functional behavior. *A primary reason for this, we believe, is because the basis functions are traditionally chosen to be circular,* though in some cases, the widths of the basis functions are adapted. While varying the width (sharpness) of the RBFs aids somewhat in improving the resolution, it still may not sufficiently help in the reduction of the number of basis functions required because many circular shaped basis functions are required to approximate sharp non-circular features.

To generalize the adaptation in the present treatment, we augment the parameter vector for each basis function with a rotation parameter vector, $\mathbf{q}$, and different spread parameters, σ_{i_k}, to control the sharpness and n-dimensional shape of each basis function, as described in Section 3.2.

$$\Theta = \left\{ w_1 \; \boldsymbol{\mu}_1^T \; \boldsymbol{\sigma}_1 \; \mathbf{q} \; \cdots \; w_h \; \boldsymbol{\mu}_h^T \; \boldsymbol{\sigma}_h \; \mathbf{q} \right\} \tag{3.53}$$

Whenever a new node or Gaussian function is added to the MMRAN network, the corresponding rotation parameters are first set to zero and the spread parameters along different directions are assumed to be equal, i.e., initially, the Gaussian functions are assumed to be circular.

The last step of the MRAN algorithm is the pruning strategy as proposed in Ref. [17]. The basic idea of the pruning strategy is to prune those nodes that contribute less than a predetermined number, δ, for S_w consecutive observations. Finally, the modified MRAN algorithm (MMRAN) can be summarized as follows:

Step 1 Compute the RBF network output using the following equation:

$$y = \sum_{i=1}^{h} w_i \Phi_i(\mathbf{x}, \boldsymbol{\Theta}) \tag{3.54}$$

$$\Phi_i(\mathbf{x}, \boldsymbol{\Theta}) = \exp\left(-\frac{1}{2}(\mathbf{x} - \boldsymbol{\mu}_i)^T \mathbf{R}^{-1}(\mathbf{x} - \boldsymbol{\mu}_i)\right) \tag{3.55}$$

Step 2 Compute different error criteria as defined in Eqs. (3.49)–(3.50).

Step 3 If all the error criteria hold, then create a new RBF center with different network parameters assigned according to the following:

$$\mathbf{w}_{h+1} = \mathbf{e}_i \tag{3.56}$$

$$\boldsymbol{\mu}_{h+1} = \mathbf{x}_i \tag{3.57}$$

$$\boldsymbol{\sigma}_{h+1_k} = \kappa \|\mathbf{x}_i - \boldsymbol{\mu}_{nearest}\|, \ \forall k = 1, 2, \cdots, n \tag{3.58}$$

$$\mathbf{q} = 0 \tag{3.59}$$

Step 4 If all criteria for adding a new node to the network are not met, then update different parameters of the network using an EKF, as described in Section 3.3.1.

Step 5 Remove those nodes of the RBF network that contribute negligibly to the output of the network for a certain number of consecutive observations.

3.5 Numerical Simulation Examples

The advantages of rotation and reshaping the Gaussian basis functions are evaluated by implementing the DCG and modified MRAN algorithm using a variety of test examples in the areas of function approximation, chaotic time series prediction and dynamical system identification problems. Most of the test case examples are either taken from the open literature or from the recently set up data modeling benchmark group [37] of the IEEE Neural Network Council. These benchmark problems are therefore a convenient framework for evaluation of accuracy, efficiency and issues affecting the relative merits of alternative learning algorithms for input-output approximation.

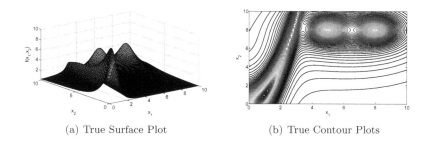

(a) True Surface Plot　　　　　　　　(b) True Contour Plots

FIGURE 3.2
True Surface and Contour Plots for Test Example 1.

In this section, we provide a comprehensive comparison of DCG and the modified MRAN algorithm with various other conventional learning algorithms. At the same time, these results also demonstrate that the inclusion of rotation and reshaping parameters significantly enhances the performance of the MRAN algorithm for all five test problems. These results will be seen as commanding evidence for the practical significance of these enhancements of the RBF input-output approximation approach.

3.5.1 Test Example 1: Function Approximation

The first test example for the function approximation is constructed by using the following analytic surface function [38]:

$$f(x_1, x_2) = \frac{10}{(x_2 - x_1^2)^2 + (1 - x_1)^2 + 1} + \frac{5}{(x_2 - 8)^2 + (5 - x_1)^2 + 1}$$
$$+ \frac{5}{(x_2 - 8)^2 + (8 - x_1)^2 + 1} \tag{3.60}$$

Figs. 3.2(a) and 3.2(b) show the true surface and contour plots of the above functional expression, respectively. According to our experience, this simple challenge function has important features, including the sharp ridge that is very difficult to learn accurately with existing function approximation algorithms using a reasonable number of nodes. To approximate the function given by Eq. (3.60), a training data set is generated by taking 10,000 uniform random samples in the interval $[0-10] \times [0-10]$ in the $X_1 - X_2$ space, while test data consists of 5,000 other uniform samples of the interval $[0 - 10] \times [0 - 10]$.

To show the effectiveness of the rotation of Gaussian basis functions, we first use the standard MRAN algorithm without the rotation parameters, as discussed in Ref. [17]. Since the performance of the MRAN algorithm depends upon the choice of various tuning parameters, several simulations were performed for various values of the tuning parameters before selecting the parameters (given in Table 3.2) which give us a compromise approximation error

TABLE 3.2

Various Tuning Parameters for MRAN and Modified MRAN Algorithms

Algori- thm	ϵ_{max}	ϵ_{min}	γ	e_{min}	$e_{r_{min}}$	κ	p_0	R	N_w	S_w	δ
Std. MRAN	3	1	0.66	0.002	0.0015	0.45	0.1	10^{-5}	200	500	0.005
Mod.- MRAN	3	1.65	0.66	0.002	0.0015	0.45	0.1	10^{-5}	200	500	0.005

for moderate dimensionality. Figs. 3.3(a) and 3.3(b) show the approximation error for the training data set and the evolution of the number of centers with number of data points. From these figures, it is clear that approximation errors are quite high even for the training data set, even though the number of Gaussian functions settled down to approximately 70 after 3,000 data points. Further, Figs. 3.3(c) and 3.3(d) show the approximated test surface and contours plots, respectively, whereas Figs. 3.3(e) and 3.3(f) show the percentage error surface and error contour plots corresponding to test data respectively. From these figures, it is apparent that approximation errors are pretty large ($\approx 15\%$) along the knife edge of the sharp ridge line while they are $< 1\%$ in other regions. Actually, the reason for the high value of the standard deviation of the approximation error for MRAN in Table 3.3 is that the much larger errors along the sharp ridge dominate the statistics. The poor performance of MRAN-type learning algorithms in this case can be attributed directly to the *inability of the prescribed circular Gaussian basis function to approximate the sharp ridge efficiently*.

Further, to show the effectiveness of the shape and rotation parameters, we modify the MRAN algorithm, as discussed in Section 3.4, by simply also including the shape and rotation parameters as adaptable while keeping the same center selection and pruning strategy. The modified MRAN algorithm is trained and tested with the same training data sets that we used for the original algorithm. In this case, too, a judicious selection of various tuning parameters is made by performing a few different preliminary simulations and selecting final tuning parameters (given in Table 3.2) which give us a near-minimum approximation error. Figs. 3.4(a) and 3.4(b) show the approximation error for the training data set and the evolution of the number of centers with the number of data points. *From these figures, it is clear that by learning the rotation parameters, the approximation errors for the training data set are reduced by almost an order of magnitude, whereas the number of Gaussian functions is reduced by half.* It should be noted, however, that a $\sim 50\%$ reduc-

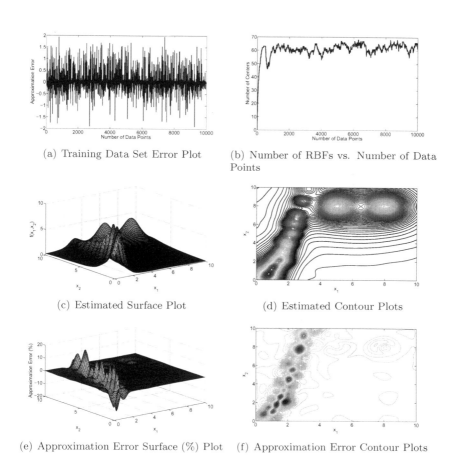

(a) Training Data Set Error Plot

(b) Number of RBFs vs. Number of Data Points

(c) Estimated Surface Plot

(d) Estimated Contour Plots

(e) Approximation Error Surface (%) Plot

(f) Approximation Error Contour Plots

FIGURE 3.3

MRAN Approximation Results for Test Example 1.

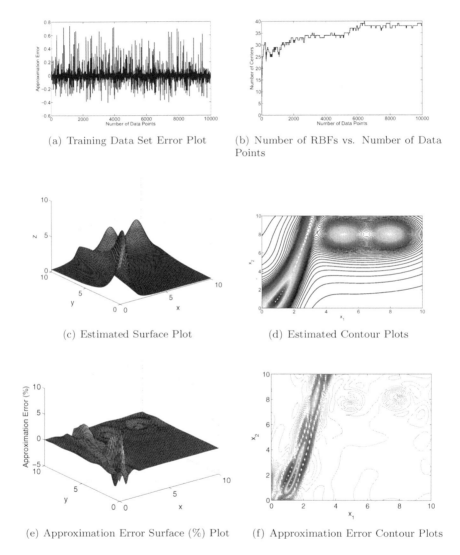

(a) Training Data Set Error Plot

(b) Number of RBFs vs. Number of Data Points

(c) Estimated Surface Plot

(d) Estimated Contour Plots

(e) Approximation Error Surface (%) Plot

(f) Approximation Error Contour Plots

FIGURE 3.4

Modified MRAN Approximation Results for Test Example 1.

tion in the number of Gaussian functions corresponds to only a 17% reduction in the number of network parameters to be learned. Figs. 3.4(c) and 3.4(d) show the approximated surface and contours plots, respectively, whereas Figs. 3.4(e) and 3.4(f) show the percentage error surface and error contour plots, respectively. As suspected, the approximation errors are significantly reduced (from over 15% to about 5%) along the knife edge of the sharp ridge line, while they are still < 1% in other regions. From Table 3.3, it is apparent that the overall mean and standard deviation of the approximation errors are also reduced very significantly.

Finally, the DCG algorithm, proposed in Section 3.3, is used to approximate the analytical function given by Eq. (3.60). As mentioned in Section 3.3, we first divide the whole input region into a total of 16 squares regions (4×4 cells); this decision was our first trial, and better results might be obtained by tuning. Then we generated a directed connectivity graph of the local maxima and minima in each sub-region that finally lead to locating and shaping only 24 radial basis functions that, after parameter optimization gave worst-case approximation errors less than 5%. This whole procedure is illustrated in Fig. 3.5.*

The DCG algorithm is also trained and tested with the same data sets that we use for the MRAN algorithm training and testing. Figs. 3.6(a) and 3.6(b) show the estimated surface and contour plots respectively for the test data. From these figures, it is clear that we are able to learn the analytical function given in Eq. (3.60) very well. In Fig. 3.6(b) the circular ($\circ$) and asterisk ($*$) marks denote the initial and final positions (after the learning process is complete) of the Gaussian centers. As expected, initially the center locations cover the global and local extremum points of the surface, and finally some of those centers, shape and rotation parameters move significantly. The optimum location, shape and orientation of those functions along the sharp ridge are critical to learn the surface accurately with a small number of basis functions. Figs. 3.6(c) and 3.6(d) show the error surface and error contour plots for the DCG approximated function. From Fig. 3.6(c), it is clear that maximum approximation error is less than 5% whereas from Fig. 3.6(d) it is clear that even though we have approximated the sharp surface very well, the largest approximation errors are still confined to the vicinity of the ridge. Clearly, we can continue introducing local functions along the ridge centered on maximum residual error locations until the residual errors are declared small enough. Already, however, advantages relative to competing methods are quite evident (the smallest approximation error and the fewest number of network parameters).

For the sake of comparison, the mean approximation error, standard deviation of approximation error and total number of network parameters learned are listed in Table 3.3 for MRAN (with and without rotation parameters)

---

*Please see the Appendix for the color versions of these figures.

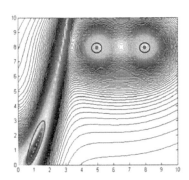

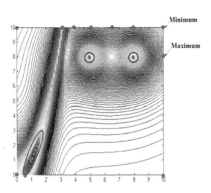

(a) Step 1: Locate Interior Extremum Points.

(b) Step 2: Locate Extremum Points along the Boundary of Input-Space.

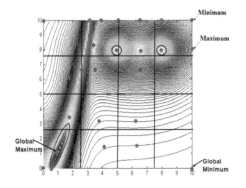

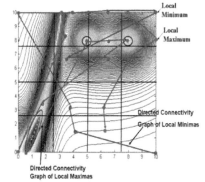

(c) Step 3: Locate Local Extremum Points.

(d) Step 4: Make a Directed Connectivity Graph of Local Extremum Points.

FIGURE 3.5

Illustration of Center Selection in the DCG Network.

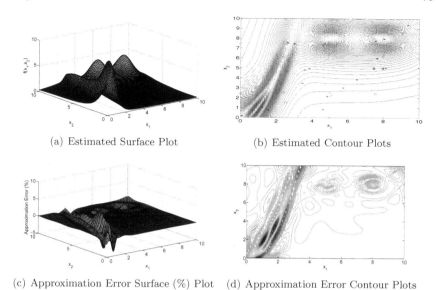

(a) Estimated Surface Plot   (b) Estimated Contour Plots

(c) Approximation Error Surface (%) Plot   (d) Approximation Error Contour Plots

FIGURE 3.6
DCG Simulation Results for Test Example 1.

and DCG algorithms. From these numbers, it is very clear that the mean approximation error and standard deviation decrease by factors from three to five if we include the rotation and shape parameters. Further, this reduction in approximation error is also accompanied by a considerable *decrease* in the number of learned parameters required to define the RBFN network in each case. It is noted that the very substantial improvement in the performance of the modified MRAN algorithm over the standard MRAN can be attributed directly to the *inclusion of shape and rotation parameters*, because the other parameter selections and learning criteria for the modified MRAN algorithm are held the same as for the original MRAN algorithm. Although there is not much difference between the modified MRAN and DCG algorithm results, in terms of accuracy, in the case of the DCG algorithm, a total of only 144 network parameters are required to be learned as compared to 232 in the case of the modified MRAN. This 33% decrease in the number of network parameters to be learned in the case of the DCG can be attributed to the *judicious selection of centers*, using the graph of maxima and minima, and the avoidance of local convergence to suboptimal values of the RBF parameters. It is anticipated that persistent optimization and pruning of the modified MRAN may lead to results comparable to the DCG results. In essence, DCG provides more nearly the global optimal location, shape and orientation parameters for the Gaussian basis functions to start the modified MRAN algorithm. For any nonlinear input-output map, our experience indicates that provision for judi-

TABLE 3.3

Comparative Results for Test Example 1

Algorithm	Mean Error	Std. Deviation (σ)	Max. Error	Number of Network Parameters
Std. MRAN	32×10^{-4}	0.1811	2.0542	280
Modified MRAN	6.02×10^{-4}	0.0603	0.7380	232
DCG	5.14×10^{-4}	0.0515	0.5475	144

cious starting parameters is vital to reliably achieving good results. The DCG algorithm brings this truth into sharp focus in the results discussed above.

3.5.2 Test Example 2: 3-Input 1-Output Continuous Function Approximation

In this section, the effectiveness of the shape and rotation parameters is shown by comparing the modified MRAN and DCG algorithms with the *Dependence Identification* (DI) algorithm [39]. The DI algorithm bears resemblance to the boolean network construction algorithms and it transforms the network training problem into a set of quadratic optimization problems that are solved by a number of linear equations. The particular test example considered here is borrowed from Ref. [39] and involves the approximation of a highly nonlinear function given by the following equation:

$$y = \frac{1}{10} \left(e^{x_1} + x_2 x_3 \cos(x_1 x_2) + x_1 x_3 \right) \qquad (3.61)$$

Here, $x_1 \in [0, 1]$ and $x_2, x_3 \in [-2, 2]$. We mention that in Ref. [17], Sundarajan et al. compared the MRAN algorithm with the DI algorithm. As in Refs. [17, 39], the input vector for MMRAN and DCG is $\mathbf{x} = \left\{ x_1 \ x_2 \ x_3 \right\}^T$, and the training data set for network learning is generated by taking 2,000 uniformly distributed random values of the input vector and calculating the associated value of y according to Eq. (3.61). Several tuning parameters for the MMRAN algorithm are given in Table 3.4. Fig. 3.7(a) shows the growth of the modified MRAN network. In the case of the DCG network, the whole input space is divided into a $2 \times 2 \times 2$ grid, thus giving us the freedom to choose the

TABLE 3.4

Various Tuning Parameters for Modified MRAN Algorithm for Test
Example 2

Algo-rithm	ϵ_{max}	ϵ_{min}	γ	e_{min}	$e_{r_{min}}$	κ	p_0	q_0	N_w	S_w	δ
Mod. MRAN	3	0.3	0.97	0.002	0.12	0.70	1	10^{-1}	10^2	2000	10^{-4}

connectivity graph of 16 centers. However, finally we settled down to a total of 4 basis functions to have mean training data set errors of the order of 10^{-3}. Further, Fig. 3.7 shows the result of testing the modified MRAN and DCG networks with the input vector **x** set to the following three parameterized functions of t, as described in Refs. [17, 39].

Test Case 1
$$x_1(t) = t$$
$$x_2(t) = 1.61 \qquad (3.62)$$
$$x_3(t) = \begin{cases} 8t - 2 & 0 \leq t < \frac{1}{2} \\ -8t + 6 & \frac{1}{2} \leq t < 1 \end{cases}$$

Test Case 2
$$x_1(t) = t$$
$$x_2(t) = \begin{cases} 8t - 2 & 0 \leq t < \frac{1}{2} \\ -8t + 6 & \frac{1}{2} \leq t < 1 \end{cases} \qquad (3.63)$$
$$x_3(t) = step(t) - 2step(t - 0.25) + 2step(t - 0.5) - \cdots$$

Test Case 3
$$x_1(t) = t$$
$$x_2(t) = step(t) - 2step(t - 0.25) + 2step(t - 0.5) - \cdots \qquad (3.64)$$
$$x_3(t) = 2\sin(4\pi t)$$

As in Refs. [17, 39], in all 3 test cases t takes on 100 evenly spaced values in the $[0, 1]$ interval. In Table 3.5, comparative results are shown in terms of percentage squared error for each test case and set of network parameters. The performance numbers for MRAN and DI algorithms are taken from Ref. [17]. From this Table and Fig. 3.7, it is clear that modified MRAN and DCG achieve smaller approximation errors with a smaller number of network parameters. Once again, the effectiveness of the shape and rotation parameters is clear from the performance difference between the standard MRAN and the modified MRAN algorithms, although the advantage is not as dramatic as in the first example.

TABLE 3.5
Comparative Results for 3-Input, 1-Output Nonlinear Function Case

Algorithm	Network Architecture	Squared Percentage Error for All Test Sets	Number of Network Parameters
Modified MRAN	3-4-1	0.0265	40
DCG	3-4-1	0.0237	40
Std. MRAN	3-9-1	0.0274	45
DI	4-280-1	0.0295	1400

3.5.3   Test Example 3: Dynamical System Identification

In this section, a nonlinear system identification problem is considered to test the effectiveness of the shape and rotation parameters. The nonlinear dynamical system is described by the following equation and is borrowed from Refs. [17, 40]:

$$y_{n+1} = \frac{1.5y_n}{1 + y_n^2} + 0.3\cos y_n + 1.2u_n \tag{3.65}$$

The particular system considered here was originally proposed by Tan et al. in Ref. [40]. In Ref. [40], a recursive RBF structure (with 42 fixed neurons and one width value (0.6391)) is used to identify the discrete-time dynamical system given by Eq. (3.65). Further, in Ref. [17] the standard MRAN algorithm is employed to predict the value of $y(n + 1)$ with 11 hidden units. It should be noted that while the number of hidden units was reduced by a factor of three, the total number of parameters (44 in the case of MRAN) to be learned was increased by 2 as compared to the total number of parameters learned in Ref. [40].

As in the previous test examples, to show the effectiveness of the shape and

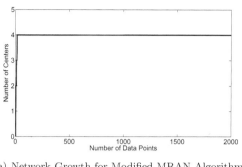

(a) Network Growth for Modified MRAN Algorithm

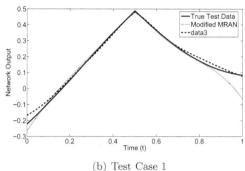

(b) Test Case 1

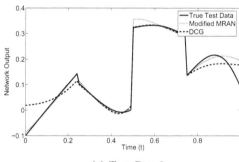

(c) Test Case 2

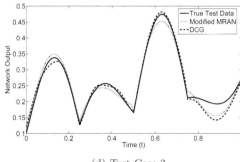

(d) Test Case 3

FIGURE 3.7

Simulation Results for Test Example 2.

rotation parameters, we first use the modified MRAN algorithm to identify the particular discrete-time system. As in Refs. [17, 40], the RBF network is trained by taking 200 uniformly distributed random samples of input signals, u_n, between -2 and 2. The network input vector, $\mathbf{x}$, is assumed to consist of y_{n-1}, and u_n, i.e.,

$$\mathbf{x} = \left\{ y_{n-1} \; u_n \right\} \tag{3.66}$$

To test the learned RBF network, test data are generated by exciting the nonlinear system by a sequence of periodic inputs [17, 40]:

$$u(n) = \begin{cases} \sin(2\pi n/250) & 0 < n \leq 500 \\ 0.8 \sin(2\pi n/250) + 0.2 \sin(2\pi n/25) & n > 500 \end{cases} \tag{3.67}$$

The different tuning parameters for the modified MRAN algorithms are given in Table 3.6. Fig. 3.8(a) shows the actual system excitation, the RBF network output learned by the modified MRAN algorithm with shape and rotation parameters, and the approximation error. Fig. 3.8(b) shows the plot of the evolution of the RBF network with the number of data points. From these plots, we can conclude that the number of hidden units required to identify the discrete-time system accurately reduces to 7 from 11 if we introduce shape and rotation optimization of the Gaussian functions in the standard MRAN algorithm. However, in terms of the total number of learning parameters, there is a reduction of only 2 parameters when we include the shape and rotation parameters in the MRAN algorithm.

Finally, the Directed Connectivity Graph learning algorithm is used to learn the unknown nonlinear behavior of the system described by Eq. (3.65). For approximation purposes, the input space is divided into a 2×2 grid, giving us the freedom to choose a maximum of 8 radial basis functions. However, the final network structure requires only 6 neurons to have approximation errors less than 5%. Fig. 3.8(c) shows the plot of the training data set approximation error with 6 basis functions, while Fig. 3.8(d) shows the actual system excitation for test data, the RBF network output learned by the DCG algorithm, and the approximation error. From these plots, we conclude that the DCG algorithm is by far the most advantageous since it requires only 6 Gaussian centers to learn the behavior of the system accurately, as compared to 42 and 11 Gaussian centers used in Refs. [40] and [17], respectively. In terms of the total number of learning parameters, the DCG algorithm is also preferable. For DCG, we need to learn only $6 \times 6 = 36$ parameters, as compared to 42 and 44 parameters for MMRAN and MRAN, respectively. This result once again reiterates our observation that the better performance of the DCG and MMRAN algorithms can be attributed to the adaptive shape and rotation learning of the Gaussian functions as well as the *judicious choice of initial centers* (in the case of DCG). It is obvious that we have achieved (*i*) more accurate convergence, (*ii*) fewer basis functions, and (*iii*) fewer network parameters, and, importantly, we have a systematic method for obtaining the starting estimates.

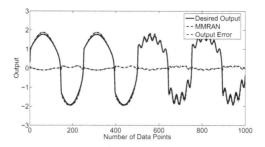

(a) Test Data Approximation Result for Modified MRAN Algorithm

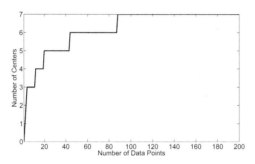

(b) Number of Centers vs. Data Points

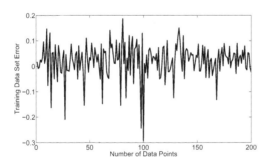

(c) Training Set Approximation Error for DCG Algorithm

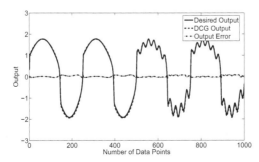

(d) Test Data Approximation Result for DCG Algorithm

FIGURE 3.8

Simulation Results for Test Example 3.

TABLE 3.6
Various Tuning Parameters for Modified MRAN Algorithm for Test
Example 3

Algorithm	ϵ_{max}	ϵ_{min}	γ	e_{min}	$e_{r_{min}}$	κ	p_0	R	N_w	S_w	δ
Mod. MRAN	3	1	0.6	0.04	0.4	0.50	1	10^{-2}	25	200	10^{-4}

3.5.4   Test Example 4: Chaotic Time Series Prediction

The effectiveness of shape and rotation parameters has also been tested with the chaotic time series generated by the Mackey-Glass time delay differential equation [41]:

$$\frac{ds(t)}{dt} = -\beta s(t) + \alpha \frac{s(t-\tau)}{1 + s^{10}(t-\tau)} \tag{3.68}$$

This equation is extensively studied in Refs. [17, 29, 42, 43] for its chaotic behavior, and is listed as one of the benchmark problems at the IEEE Neural Network Council website [37]. To compare directly to the previous studies [17, 29, 42, 43], we choose the same parameter values: $\alpha = 0.2$, $\beta = 0.1$, $\tau = 17$, and $s(0) = 1.2$. Further, to generate the training and testing data set, the time series Eq. (3.68) is integrated by using the fourth-order Runge-Kutta method to find the numerical solution. This data set can be found in the file mgdata.dat belonging to the fuzzy logic toolbox of MATLAB 7 and at the IEEE Neural Network Council website [37].

Once again, to study the effectiveness of introducing shape and rotation parameters only, we used the modified MRAN algorithm and the DCG algorithm to perform a short-term prediction of this chaotic time series. We predict the value of $s(t+6)$ from the current value of $s(t)$ and the past values of $s(t-6)$, $s(t-12)$ and $s(t-18)$. As in previous studies [17, 29, 42, 43], the first 500 data set values are used for network training while the remaining 500 values are used for testing purposes. The different tuning parameters for the modified MRAN algorithm are given in Table 3.7. For the DCG approximation purposes, the input space is divided into a $2 \times 2 \times 2 \times 2$ grid, giving us the freedom to choose a maximum of 32 radial basis functions. However, the final network structure required only 4 neurons to achieve approximation errors less than 5%. We mention that due to the availability of a small number of training data set examples, we used the Levenberg-Marquardt [2] algorithm to efficiently optimize the DCG network.

TABLE 3.7
Various Tuning Parameters for Modified MRAN Algorithm for Test
Example 4

Algo-rithm	ϵ_{max}	ϵ_{min}	γ	e_{min}	$e_{r_{min}}$	κ	p_0	R	N_w	S_w	δ
Mod. MRAN	2	0.5	0.66	10^{-5}	10^{-4}	0.27	1	10^{-1}	10^2	10^3	10^{-4}

Fig. 3.9(a)* shows the MMRAN network growth with the number of training
data set examples, while Figs. 3.9(b)* and 3.9(c)* show the plots for approximated test data and approximation test data error, respectively. From these
plots, we can conclude that the MMRAN algorithm requires only 6 Gaussian
centers to learn the behavior of the system accurately as compared to 29 and
81 Gaussian centers used in Refs. [17] and [29], respectively. In terms of the
total number of learning parameters, the MMRAN algorithm is also preferable, as compared to the MRAN and RAN algorithms. For the MMRAN
algorithm, we need to learn only $6 \times 15 = 90$ parameters, as compared to 174
parameters required for the MRAN algorithm. In the case of the DCG algorithm, the number of Gaussian centers required was reduced even further to
only 4, while the total number of learned parameters reduced to 60, as compared to 90 in the case of the MMRAN algorithm and 174 for the standard
MRAN algorithm. In Ref. [43], Table IX compares the various algorithms
presented in the literature in terms of their root mean squared error (RMSE)
for this particular problem. Here, in Table 3.8, we present comparative results for MMRAN, DCG and five other algorithms. The direct comparison
of MRAN and MMRAN results reveals the fact that inclusion of the shape
and rotation parameters greatly enhances the approximation accuracy, while
significantly reducing the number of parameters required to define the RBF
network for a *particular algorithm*. It should also be noted that both the
DCG and MMRAN algorithms performed very well as compared to all other
algorithms for this particular example, in terms of both smallness of the RMS
error and the number of free network parameters.

---

*Please see the Appendix for the color versions of these figures.

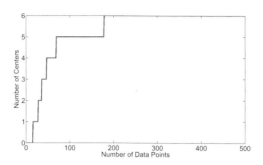

(a) Number of Centers vs. Data Points

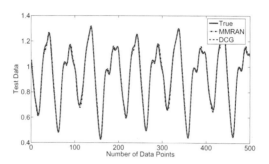

(b) Test Data Approximation Result

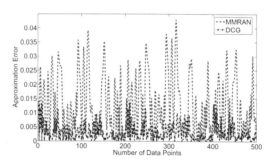

(c) Test Data Approximation Error

FIGURE 3.9
Simulation Results for Test Example 4.

TABLE 3.8

Comparative Results for Mackey-Glass Chaotic Time Series Prediction Problem

Algorithm	Network Architecture	RMS Error	Number of Network Parameters
MRAN	4-29-1	0.035	174
Modified MRAN	4-6-1	0.0164	90
DCG	4-4-1	0.004	60
Genetic Algorithm + Fuzzy Logic [43]	$9 \times 9 \times 9 \times 9$	0.0379	6633
Pomares 2000 [44]	$3 \times 3 \times 3 \times 3$	0.0058	101
Pomares 2003 [43]	4-14-1	0.0045	84
Pomares 2003 [43]	4-20-1	0.0029	120

TABLE 3.9
Various Tuning Parameters for Modified MRAN Algorithm for Test
Example 5

Algo-rithm	ϵ_{max}	ϵ_{min}	γ	e_{min}	$e_{r_{min}}$	κ	p_0	R	N_w	S_w	δ
Mod. MRAN	2	0.9	0.99	10^{-2}	10^{-2}	0.7	1	1	500	5000	10^{-4}

3.5.5 Test Example 5: Benchmark Against the On-Line Structural Adaptive Hybrid Learning (ONSAHL) Algorithm

In this section, we present a comparison of the MMRAN and DCG algorithms with the On-line Structural Adaptive Hybrid Learning (ONSAHL) algorithm on a nonlinear system identification problem from Ref. [28]. The ONSAHL algorithm uses a Direct Linear Feedthrough Radial Basis Function (DLF-RBF) network and an error-sensitive cluster algorithm to automatically determine the number of RBF neurons, and to adapt their center positions, their widths, and the output layer weights. This algorithm, however, does not include shape and rotation parameters. The nonlinear dynamical system is described by the following difference equation and is borrowed from Ref. [28]:

$$y(n) = \frac{29}{40} \sin \left(\frac{16u(n-1) + 8y(n-1)}{3 + 4u(n-1)^2 + 4y(n-1)^2} \right)$$
$$+ \frac{2}{10} \left(u(n-1) + y(n-1) \right) + \epsilon(n) \tag{3.69}$$

As in Ref. [28], $\epsilon(n)$ denotes a Gaussian white noise sequence with zero mean and a variance of 0.0093. A random signal uniformly distributed in the interval $[-1, 1]$ is used for the excitation $u(n)$ in the system of Eq. (3.69). The network input vector $\mathbf{x}$ is assumed to consist of $y(n-1)$ and $u(n-1)$, while the network output vector consists of $y(n)$. Eq. (3.69) is simulated with zero initial conditions to generate response data for 10,000 integer time steps. Out of these 10,000 data points, the first 5,000 are used for training purposes, while the remaining 5,000 points are used for testing purposes. Fig. 3.10(a)* shows the plot of true test data.

In this case, several MRAN tuning parameters are given in Table 3.9. For DCG approximation purposes, the input space is divided into a 2×2 grid,

---

*Please see the Appendix for the color version of this figure.

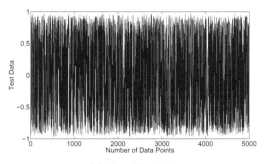

(a) True Test Data

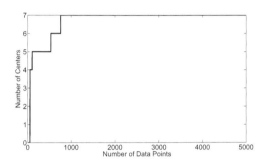

(b) Number of Centers vs. Data Points

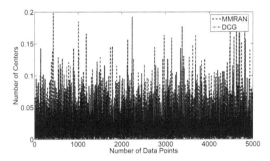

(c) Absolute Test Data Set Approximation Error

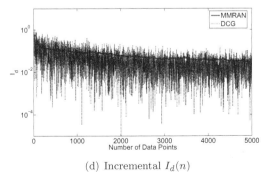

(d) Incremental $I_d(n)$

FIGURE 3.10
Simulation Results for Test Example 5.

TABLE 3.10
Comparative Results for Test Example 5

Algorithm	Network Architecture	Mean $I_d(n)$	Number of Network Parameters
Modified MRAN	2-7-1	0.0260	42
DCG	2-6-1	0.0209	36
Std. MRAN	2-11-1	0.0489	44
ONSAHL	2-23-1	0.0539	115

giving us the freedom to choose a maximum of 8 radial basis functions. However, the final network structure consists of only 6 neurons in order to have approximation errors less than 5%. For comparison purposes, we also use the same error criteria as defined in Ref. [28].

$$I_d(n) = \frac{1}{50} \sum_{j=0}^{49} |y(n-j) - \hat{y}(n-j)| \qquad (3.70)$$

Fig. 3.10(b)* shows the plot of the MMRAN network growth with the number of training data points, while Figs. 3.10(c)* and 3.10(d)* show the plot of absolute approximation error and incremental $I_d(n)$, respectively. In Ref. [17], the standard MRAN algorithm is employed for system identification purposes using 11 neurons while the ONSAHL algorithm is employed using 23 neurons. From the results presented in Ref. [17], it is clear that the MRAN algorithm uses a smaller number of neurons as compared to the ONSAHL algorithm to accurately represent the given dynamical system. From Fig. 3.10(b), it is clear that the number of neurons required to identify the discrete-time system accurately further reduces to 7 from 11 if shape and rotation adaptation of the Gaussian RBF is incorporated in the MRAN algorithm. However, in terms of the total number of learning parameters, there is a reduction of only 2 parameters if we include the shape and rotation parameters in the MRAN algorithm.

---

*Please see the Appendix for the color versions of these figures.

From these plots, we can also conclude that the DCG algorithm requires only 6 Gaussian centers to learn the behavior of the system accurately, as compared to 23 and 11 Gaussian centers used in Refs. [28] and [17], respectively. In terms of the total number of learning parameters, the DCG algorithm is again preferable. For DCG, we need to learn only 36 parameters as compared to 42 and 44 parameters for MMRAN and MRAN, respectively, while substantially reducing the approximation error. Finally, Table 3.10 summarizes the comparison of results in terms of approximation error and the number of free network parameters. These results once again reiterate our observation and support the conclusion that the better performance of the DCG and MMRAN algorithms can be attributed to the inclusion of shape and rotation optimization of Gaussian functions, as well as the optimization of their centers and fully parameterized shape and spread. These advantages, taken with the previous four problem results, provide compelling evidence for the merits of the shape and rotation optimization of the Gaussian basis functions, as well as a directed connectivity graph algorithm to initialize estimates for these parameters.

3.6   Summary

A direction-dependent RBFN learning algorithm has been discussed to obtain a minimal RBF network. New approaches are introduced and tested on a variety of examples from a variety of disciplines, such as continuous function approximation, dynamic system modeling and system identification, nonlinear signal processing, and time series prediction. In all of these diverse test problems, the two new algorithms introduced are found to produce more compact RBF networks with the same or smaller errors compared to the existing methods known to have been treated on the same benchmark problems. The results are of direct utility in addressing the "curse of dimensionality," accuracy of convergence, and frequent redundancy of neural network approximation.

The results presented in this chapter serve to illustrate the usefulness of shape and rotation optimization of the Gaussian basis functions, as well as a directed connectivity graph algorithm to initialize estimates for these parameters. The shape and rotation optimization of the Gaussian functions not only helps us in approximating the complex surfaces better, but also helps in greatly reducing the numbers of hidden units. We believe that basis function shape and rotation optimization can be incorporated into many existing learning algorithms to very significantly enhance performance without much difficulty. This fact was illustrated by our modification of an existing MRAN learning algorithm. However, much research is required to extend and optimize the methodology for general multi-resolution approximations in high-dimensional spaces. We fully appreciate that results from any finite set of tests are difficult to extrapolate; however, testing the new RBF algorithms on five benchmark problems and providing comparisons to the most obvious competing algorithms does provide some compelling evidence and a basis for optimism as regards future applications. Finally, we mention that proving the minimality of the RBF network (using any learning algorithm for that matter) is an open problem in the field of approximation theory. The word "minimal" in the chapter only signifies that we have sought a minimum parameter representation, and no more compact networks are known to exist in the literature for all of the test problems and the test data considered in this chapter. However, one can use sparse approximation techniques, like Support Vector Machines (SVMs) [45, 46], to find the minimal RBF network from a given RBF network, which is optimal in the sense of the trade-off between approximation error and network size (measured in $\mathcal{L}_1$ norm).

We mention that this chapter has concentrated on improved methods for nonlinear parameterization of input-output maps. Subsequent chapters introduce novel and powerful new methods for linearly parameterizing input-output maps; these new methods have exceptional potential for high-dimensioned, multi-scaled input-output maps.

4

Multi-Resolution Approximation Methods

A mathematician is a device for turning coffee into theorems.

Paul Erdos

4.1 Introduction

In the previous chapter, we showed that the learning of shape and orientation parameters of a basis function significantly improves the approximation
capability of a Gaussian basis function. This intuitively comfortable fact was
illustrated by considering a variety of examples from a variety of disciplines
such as continuous function approximation, dynamic system modeling and
system identification, nonlinear signal processing, and time series prediction.
Although the RBF learning algorithms presented in Chapter 3 are shown to
work very well for different test examples, there remain several challenging
issues concerning the complexity and convergence of the RBF model. The
nonlinear RBF model is global. This has both advantages and disadvantages,
but ultimately for very high-dimensioned problems, it is likely defeated by
the curse of high dimensionality (computation burden and convergence difficulties, mainly). Although successes have been many, the computational cost
associated with learning these parameters and the convergence of nonlinear
estimation algorithms remain obstacles that limit applicability to problems of
low to moderate dimensionality.

A key question regarding the proper selection of an approximation algorithm is, "How irregular is the input-output map?" Qualitatively, global
best fit of the input-output map should be sufficient if the slope of the input-
output map is smooth globally without large local variations in the space-time
frequency context. In the presence of irregular localized features, a multi-
resolution learning algorithm may be required to take care of local as well as
global complexity of the input-output map.

As the name suggests, multi-resolution approximation can be defined as a
mathematical process of hierarchically decomposing the input-output approximation to capture both macroscopic and microscopic features of the system
behavior.

The unknown function underlying any given measured input-output data can be considered as consisting of high frequency local input-output variation details superimposed on the comparatively low frequency smooth background. More than two levels of granularity (resolution) will be required in a general setting. The term "resolution" can be defined as the scale to measure the details of the input-output data that cannot be discerned. At a given resolution, the input-output data is approximated by ignoring all variations below that scale [47]. At each stage, finer details are added to the coarser description, providing a successively better approximation to the input-output data. Eventually, when the resolution goes to infinity, we would expect the approximate mathematical model to approach the exact smooth function underlying any given input-output data. In rigorous mathematical terms, we can view the space of functions that are square integrable as composed of a sequence of two orthogonal subspaces W_k and V_j such that the approximation at resolution level j is in V_j, and the higher frequency details are in W_k. This brings us to the following formal definition of the multi-resolution approximation:

A sequence $\{V_j\}_{j \in \mathbb{Z}}$ of closed subspaces is a *multi-resolution approximation* if the following 6 properties are satisfied:

1. $\forall j \in \mathbb{Z}, \; V_j \subset V_{j+1}$

2. $\displaystyle\lim_{j \to -\infty} V_j = \cap_{-\infty}^{\infty} V_j = \{0\}$

3. $\displaystyle\lim_{j \to \infty} V_j = \overline{(\cup_{-\infty}^{\infty} V_j)} = \mathcal{L}^2(\mathbb{R})$

4. $\forall j \in \mathbb{Z}, \; f(x) \in V_j \Leftrightarrow f(2x) \in V_{j+1}$

5. $\forall \, (k) \in \mathbb{Z}, \; f(x) \in V_0 \Leftrightarrow f(x - k) \in V_0$

6. There exists a function $\theta(x)$, called the scaling function, such that $\{\theta(x - k)\}$ is an orthonormal basis of V_0

where $\mathbb{Z}$ denotes an index set for resolution index j, $\cap_{-\infty}^{\infty} V_j$ denotes the intersection of all possible subspaces V_j and $\overline{(\cup_{-\infty}^{\infty} V_j)}$ represents the closure of the union of all possible subspaces V_j.

According to the aforementioned definition of Multi-Resolution Approximation (MRA), the starting point for the MRA analysis is the decomposition of function space, V_0, into a sequence of subspaces, V_j. Now, the first condition implies that the subspace V_j be contained in all the higher subspaces. Literally, it means that information contained at level j must be included in the information at a higher resolution, which is a reasonable requirement. The second condition corresponds to the fact that as resolution gets coarser and coarser, the approximation becomes more crude, and in the limit $j \to -\infty$, we should get a constant function which can only be a zero function due to the square integrable constraint. The third condition is the opposite of the second condition and states that as the resolution is increased, more details are included in the approximation, and in the limit $j \to \infty$, we should get back the entire space $\mathcal{L}^2(\mathbb{R})$. The fourth condition is equivalent to scale or dilation invariance of space V_j while the fifth condition corresponds to translation and dilation invariance. The sixth and final condition guarantees the existence of a orthonormal basis for V_j. Note if $\theta(t-k)$ forms an orthonormal basis for V_0 then by scale and translation invariance $\theta_{jk} = 2^{\frac{j}{2}}\theta(2^j t - k)$ forms an orthonormal basis for V_j [47, 48].

Conventional spline, piecewise linear approximation, and wavelet approximation are some examples of *multi-resolution analysis*. In this chapter, we briefly discuss some of the most popular multi-resolution algorithms.

4.2 Wavelets

The term "multi-resolution" enjoys wide popular use in wavelet analysis as wavelets allow a function to be described in terms of a coarse overall shape, plus details that range from broad to narrow [48]. In fact, the definition of multi-resolution analysis was first coined during the development of wavelet analysis. In this section, we briefly discuss wavelet analysis; readers should refer to Refs. [48–50] for more details on this topic.

Wavelets constitute a special set of basis functions for the square-integrable space $\mathcal{L}^2(\mathbb{R})$ which is generated from a single basis function known as a *mother wavelet* by the translation and dilation process. As will be evident, there is no unique mother wavelet, and each choice affects the details of the wavelet approximation that flow from selecting the mother wavelet. To explain it further, from the definition of multi-resolution analysis, we consider a sequence of subspaces V_j such that $\bigcup_j V_j = \mathcal{L}^2(\mathbb{R})$ and $V_j \subset V_{j+1}$. Now, let us decompose the space V_{j+1} into two orthogonal subspaces V_j and W_j:

$$V_{j+1} = V_j \oplus W_j, \quad V_j \perp W_j \tag{4.1}$$

By iterating Eq. (4.1) and making use of the second condition of MRA, we can write

$$V_{j+1} = W_j \oplus W_{j-1} \oplus W_{j-2} \oplus \cdots = \bigoplus_{i=-\infty}^{j} W_k, \quad \mathcal{L}^2(\mathbb{R}) = \bigoplus_{i=-\infty}^{\infty} W_k \tag{4.2}$$

Further, note that all W_j are orthogonal to each other since $W_j \perp V_j$ and thus W_j should be orthogonal to any subspace of V_j, including W_{j-1}. Hence, the set of W_j serves as a basis for the square-integrable function space $\mathcal{L}^2(\mathbb{R})$.

Now, let us assume that $\{\phi_{jk}(x)\}$ are the basis functions for each subspace V_j. Further, from the first and sixth conditions of multi-resolution analysis, we can write scaling function $\theta(x)$ as a linear combination of ϕ_{1k}, i.e., basis functions for V_1.

$$\theta(x) = \sum_k c_k \phi_{1k}(x) \tag{4.3}$$

Now, using the scale (fourth) and translation (fifth) invariance conditions, $\sqrt{2}\theta(2x-k)$ is an orthonormal basis for V_1, and in particular $\phi_{jk} = 2^{\frac{j}{2}}\theta(2^j x - k)$ is an orthonormal basis for V_j. Thus, Eq. (4.3) can be rewritten as

$$\theta(x) = \sum_k c_k \phi_{1k}(x) = \sqrt{2} \sum_k c_k \theta(2x - k) \tag{4.4}$$

and c_k can be computed as

$$c_k = \int\limits_{-\infty}^{\infty} \theta(x)\theta(2x - k)dx \tag{4.5}$$

Note that Eq. (4.5) relates basis functions at two different resolution levels.

Now, let us consider a function $\vartheta(x) \in W_0$ and making use of the fact that $W_0 \subset V_1$, we can write $\vartheta(x)$ as a linear combination of $\phi_{1k}(x)$

$$\vartheta(x) = \sum_k d_k \phi_{1k}(x) = \sqrt{2} \sum_k d_k \theta(2x - k) \tag{4.6}$$

where d_k can be computed as

$$d_k = \int\limits_{-\infty}^{\infty} \vartheta(x)\phi(2x - k)dx \tag{4.7}$$

Note that Eq. (4.6) relates function $\vartheta(x) \in W_0 \subset V_1$ to scaling function $\theta(x) \in V_0$. This is known as the *wavelet equation*. Further, from the definition of multi-resolution analysis and Eqs. (4.1) and (4.6), we infer that $\{\varphi(x)_{jk} = 2^{\frac{j}{2}}\vartheta(2^j x - k)\}_k$ forms an orthonormal basis for W_j. Hence, an orthonormal basis $\{\varphi(x)_{jk} = 2^{\frac{j}{2}}\vartheta(2^j x - k)\}_{jk}$ for $\mathcal{L}^2(\mathbb{R})$ can be constructed from a single function $\vartheta(x)$, known as the *mother wavelet*. So, one can construct an orthonormal basis for $\mathcal{L}^2(\mathbb{R})$ from a single function $\theta(x)$ associated with a multi-resolution analysis. In summary, the problem of finding a wavelet basis corresponds to finding a function $\theta(x)$ that satisfies all the necessary conditions of multi-resolution analysis and satisfies Eq. (4.4). Clearly, different choices for scaling function $\theta(x)$ will lead to different wavelet bases.

To illustrate the procedure of constructing a mother wavelet function, $\vartheta(x)$, let V_j be the space of all square-integrable functions $f(x)$ such that the Fourier transform of $f(x)$, $F(\omega) = 0$, $\omega > (j-1)\pi$. From Shannon's sampling theorem [20], $\theta(x-k) = \frac{\sin \pi(x-k)}{\pi(x-k)}$ forms an orthonormal basis for V_0.

$$f(x) = \sum_k F(n) \frac{\sin \pi(x - k)}{\pi(x - k)} \tag{4.8}$$

Now, note that $f(2x)$ can be written as follows:

$$h(x) = f(2x) = \sum_k H\left(\frac{n}{2}\right) \frac{\sin \pi \left(2x - k\right)}{\pi \left(2x - k\right)} \tag{4.9}$$

and, according to Shannon's sampling theorem, the Fourier transform of $h(x)$, $H(\omega) = 0$, $\omega > 2\pi$ which means that $h(x) \in V_1$. Also, it is clear that $V_0 \subset V_1$. Similarly, we can show that in general $V_j \subset V_{j+1}$, and if $g(x) \in V_j$, then $g(2x) \in V_{j+1}$. So, we can conclude that the sequence of subspaces V_j forms a multi-resolution analysis on $\mathcal{L}^2(\mathbb{R})$ with scaling function $\theta(x)$ given by the following equation:

$$\theta(x) = \frac{\sin \pi x}{\pi x} \tag{4.10}$$

Now, to find the expression for the mother wavelet, $\vartheta(x)$, let us consider the orthogonality condition of $\theta(x - k)$ and $\vartheta(x - k)$:

$$\int_{-\infty}^{\infty} \theta(x) \bar{\vartheta}(x - k) dx = 0 \tag{4.11}$$

Taking the Fourier transformation on both sides of the above equation leads to [20]

$$\frac{1}{2\pi} \int_{-\infty}^{\infty} \Theta(\omega) \bar{\vartheta}(\omega) e^{i\omega k} d\omega = \frac{1}{2\pi} \int_{0}^{2\pi} \sum_n \Theta(\omega + 2n\pi) \bar{\vartheta}(\omega + 2n\pi) e^{i\omega k} d\omega = 0$$

where

$$\Theta(\omega) = \int_{-\infty}^{\infty} \theta(x) e^{-i\omega t} dt \tag{4.12a}$$

$$\vartheta(\omega) = \int_{-\infty}^{\infty} \vartheta(x) e^{-i\omega t} dt \tag{4.12b}$$

Hence, we can conclude that

$$\sum_n \Theta(\omega + 2n\pi) \bar{\vartheta}(\omega + 2n\pi) = 0 \tag{4.13}$$

Further, it is easy to show that Eq. (4.4) in frequency space can be written as

$$\Theta(\omega) = \left(\sum_k \frac{c_k}{\sqrt{2}} e^{-ik\frac{\omega}{2}}\right) \Theta\left(\frac{\omega}{2}\right) = C\left(\frac{\omega}{2}\right) \Theta\left(\frac{\omega}{2}\right) \tag{4.14}$$

From Eq. (4.10), it is easy to check that

$$\Theta\left(\frac{\omega}{2}\right) = \begin{array}{ll} 1 & \text{for } \omega \in [-2\pi, 2\pi] \\ 0 & \text{elsewhere} \end{array} \tag{4.15}$$

From Eqs. (4.14) and (4.20), and making use of the fact that $C(\omega)$ is 2π periodic, we can write

$$C(\omega) = \sum_n \Theta\left(\omega + 2n\pi\right) \tag{4.16}$$

Further, the Fourier transform of Eq. (4.6) leads to

$$\vartheta(\omega) = \left(\sum_k \frac{d_k}{\sqrt{2}} e^{-ik\frac{\omega}{2}}\right) \Theta\left(\frac{\omega}{2}\right) = D\left(\frac{\omega}{2}\right) \Theta\left(\frac{\omega}{2}\right) \tag{4.17}$$

Now, the substitution of Eqs. (4.14) and (4.17) in Eq. (4.13) leads to

$$C(\omega)D(\omega) + C(\omega + \pi)\bar{D}(\omega + \pi) = 0 \tag{4.18}$$

Note that the above equation is satisfied if $D(\omega)$ is chosen to be

$$D(\omega) = -e^{-i\omega}\bar{C}(\omega + \pi) \tag{4.19}$$

From Eqs. (4.13), (4.16) and (4.19), we have

$$\vartheta(\omega) = \begin{array}{ll} e^{-i\frac{\omega}{2}} & \text{for } \omega \in [-\pi, \pi] \\ 0 & \text{elsewhere} \end{array} \tag{4.20}$$

Finally, the inverse Fourier transform of $\vartheta(\omega)$ leads to the following expression for mother wavelet $\vartheta(x)$, known as the *Shannon wavelet*:

$$\vartheta(x) = \frac{\sin \pi \left(x - \frac{1}{2}\right) - \sin 2\pi \left(x - \frac{1}{2}\right)}{\pi \left(x - \frac{1}{2}\right)} \tag{4.21}$$

Fig. 4.1 shows the plot of the *Shannon wavelet*, and further, the orthonormal basis functions corresponding to the Shannon wavelet can be constructed as follows:

$$\{\varphi(x)_{jk} = 2^{\frac{j}{2}}\vartheta(2^j x - k)\}_{jk} \tag{4.22}$$

Now, one can approximate any smooth functions as a linear combination of wavelet basis functions and can use least squares methods to find unknown amplitude corresponding to various resolution levels. We demonstrate this process in Example 4.1.

Example 4.1
Let us consider the problem of approximating the following smooth function using Shannon wavelet basis functions of Eq. (4.22) over a compact interval $[-5, 5]$ using the least squares method.

$$f(x) = 1.1(1 - x - 2x^2)e^{-\frac{x^2}{2}} \tag{4.23}$$

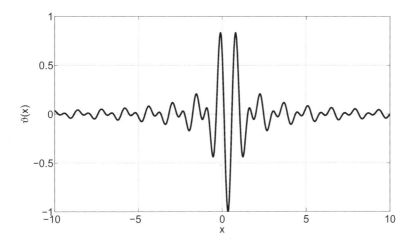

FIGURE 4.1
Shannon Mother Wavelet, $\vartheta(x)$.

We assume that we have 5,000 pairs of observations $(x_1, y_1), \cdots, (x_{5000}, y_{5000})$ of (x, y), and that a suitable model for the generation of the data is

$$y_i = f(x_i) + \nu_i, \quad i = 1, \cdots, m \qquad (4.24)$$

where ν_i denotes random white noise of standard deviation of 0.01. We assume that the approximation of $f(x)$, $\hat{f}(x)$ can be written as a linear combination of linearly independent Shannon wavelet basis functions of Eq. (4.22):

$$\hat{f}(x) = \sum_{j=0}^{n} \sum_{k=0}^{m} c_{jk} \varphi_{jk}(x) \qquad (4.25)$$

To study the effect of the number of basis functions on the approximation accuracy, we vary the resolution level n and translational level m. Fig. 4.2 shows the plot of true and approximated signals using various orders of wavelet basis functions. As expected, the approximation error decreases as we increase either n or m. Note that these results are much better than those obtained by polynomial basis functions in Chapter 1. This improvement can be attributed to the multi-resolution property of Shannon wavelet basis functions. $\quad$ ▯

Example 4.2
Let us consider the problem of approximating the following analytic surface

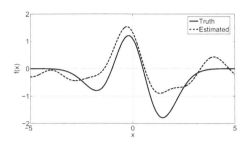

(a) $n = 1$, $m = 2$

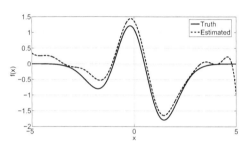

(b) $n = 1$, $m = 5$

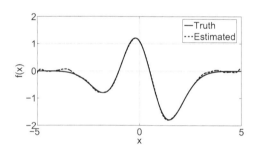

(c) $n = 1$, $m = 10$

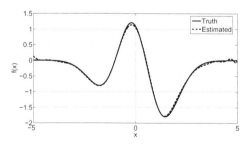

(d) $n = 2$, $m = 10$

FIGURE 4.2

Least Squares Approximation Using Wavelet Basis Functions of Various Orders.

function introduced in Chapter 3.

$$f(x_1, x_2) = \frac{10}{(x_2 - x_1^2)^2 + (1 - x_1)^2 + 1} + \frac{5}{(x_2 - 8)^2 + (5 - x_1)^2 + 1}$$
$$+ \frac{5}{(x_2 - 8)^2 + (8 - x_1)^2 + 1} \tag{4.26}$$

Figs. 3.2(a) and 3.2(b) show the true surface and contour plots of the above functional expression, respectively. As mentioned earlier, this simple challenge function has important features, including the sharp ridge that is very difficult to learn accurately with global function approximation algorithms with a reasonable number of basis functions.

For simulation purposes, a training data set is generated by taking 10,000 uniform random samples in the interval $[0 - 10] \times [0 - 10]$ in the $x_1 - x_2$ space, while test data consist of 5,000 other uniform samples of the interval $[0 - 10] \times [0 - 10]$.

We assume that the approximated surface $\hat{f}(x_1, x_2)$ can be written as a linear combination of 2-D Shannon wavelet basis functions which are constructed by the tensor product of 1-D wavelet basis functions:

$$\hat{f}(x_1, x_2) = \sum_{j=0}^{n} \sum_{k=0}^{m} c_{jk} \varphi_{jk}(x_1, x_2)$$

where

$$\varphi_{jk}(x_1, x_2) = \varphi_{j_1 k_1}(x_1) \varphi_{j_2 k_2}(x_2), \ j_1 + j_2 = j, \ \& \ k_1 + k_2 = k \tag{4.27}$$

Further, we vary the resolution level n and translational level m. We mention that 2-D wavelet basis functions were translated over a 20×20 grid over the entire $x_1 - x_2$ space, and dilation level n is varied from 1 to 2. Fig. 4.3 shows the plot of true and approximated surface and corresponding approximation error surfaces using various orders of wavelet basis functions. As expected, the approximation error decreases as we increase n. We mention that these MRA linear results are comparable in accuracy to what we obtained by the nonlinear RBF approximations in Chapter 3. □

Finally, we mention that generally it is not possible to find an analytical expression for wavelet basis functions corresponding to any given scaling functions with regard to a multi-resolution analysis. Further, there is no guarantee that resulting wavelets will be compactly supported. However, in Ref. [49], Daubechies develops a new method to find a closed-form analytic expression for a class of compactly supported orthogonal wavelets. This development is outside the scope of this presentation. Interested readers should refer to Ref. [49] for details on this method.

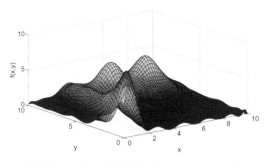

(a) Approximated Surface ($n = 1$, $m = 400$)

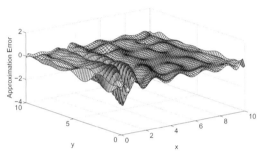

(b) Error Surface ($n = 1$, $m = 400$)

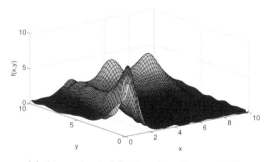

(c) Approximated Surface ($n = 2$, $m = 400$)

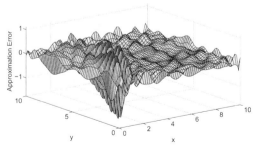

(d) Approximated Surface ($n = 2$, $m = 400$)

FIGURE 4.3
Least Squares-Based Wavelet Approximation of Analytical Surface Given by
Eq. (4.26).

4.3 Bèzier Spline

Bèzier splines or B-splines are piecewise continuous polynomial curves where each of the polynomials is in Bèzier form. These were first introduced by Pierre Bèzier for automobile design in 1969. In this section, we briefly discuss input-output approximation using B-splines. The reader should refer to Refs. [51,52] for more details.

Like any conventional spline curve, each segment of B-splines is an nth order polynomial curve defined as

$$P(x) = \sum_{i=0}^{n} p_i B_i^n(x) \tag{4.28}$$

where P_i are the control points and $B_i^n(x)$ are the nth order Bernstein polynomials given by

$$B_i^n(x) = {}^nC_i x^i (1-x)^{n-i}, \; i = 0, 1, \cdots, n, \; 0 \le x \le 1 \tag{4.29}$$

We mention that unlike conventional spline curves, control points p_i just control the continuity of the curve, and the Bèzier curve does not generally pass through them. However, Bèzier curves pass through their end points due to the following remarkable properties of Bernstein polynomials:

$$B_i^n(x) \ge 0, \; \forall x \in [0, 1] \tag{4.30}$$

$$\sum_{i=0}^{n} B_i^n(x) = 1, \; \forall x \in [0, 1] \tag{4.31}$$

$$B_i^n(0) = B_i^n(1) = 0, \; i = 1, 2, \cdots, n-1, \; \forall x \in [0, 1] \tag{4.32}$$

$$B_0^n(0) = B_n^n(1) = 1, \; \forall x \in [0, 1] \tag{4.33}$$

Further, differentiating both sides of Eq. (4.29) leads to

$$\frac{dB_i^n(x)}{dx} = {}^nC_i x^{i-1} x^{n-i-1} \left[(n-i)t + i(1-t) \right]$$
$$= n \left[B_{i-1}^{n-i}(x) - B_i^{n-1}(x) \right] \tag{4.34}$$

Now, differentiating Eq. (4.28) with respect to x and making use of Eq. (4.34), we get

$$\frac{dP(x)}{dx} = P'(x) = \sum_{i=0}^{n} p_i \frac{d}{dx} B_i^n(x) \tag{4.35}$$

Evaluating the above equation at end points of the curve, i.e., for $x = 0$ and $x = 1$, and making use of Eqs. (4.30)–(4.33), we get

$$P'(0) = n(P_1 - P_0), \; P'(1) = n(P_n - P_{n-1}) \tag{4.36}$$

Similarly, we can compute the second derivative of $P(x)$ at end points as follows:

$$P''(0) = n(n-1)(P_2 - 2P_1 + P_0), \ P''(1) = n(n-1)(P_n - 2P_{n-1} + P_{n-2}) \quad (4.37)$$

Like any polynomial curve, one needs a very high order Bèzier curve to accurately approximate a complex map. To alleviate this problem, a piecewise curve can be generated while taking care of smoothness constraints at the boundary. This results in a curve which is independent of the number of control points, for example, if we construct a second order B-spline curve using two cubic Bèzier segments, $P_1(x)$ and $P_2(x)$. Note that one needs a total of 8 control points (4 for $P_1(x)$ and 4 for $P_2(x)$) to construct these two segments independently. If we want the final curve to be second order smooth, one needs to satisfy the following constraints:

$$P_1(1) = P_2(0) \quad (4.38)$$
$$P_1'(1) = P_2'(0) \quad (4.39)$$
$$P_1''(1) = P_2''(0) \quad (4.40)$$

As a consequence of these constraints, one has only 5 degrees of freedom instead of 8. In general, a B-spline with $n + 1$ control points consists of $n - k + 1$ Bèzier segments, each of degree k. It can be easily shown that a B-spline of degree $k - 1$ is given by

$$f(x) = \sum_{i=0}^{n} p_i N_i^k(x), \ x_{k-1} \le x \le x_{n+1} \quad (4.41)$$

where p_i are the control points and $N_i^k(x)$ are B-spline basis functions defined as follows:

$$N_i^k(x) = \frac{x - x_i}{x_{i+k-1} - x_i} N_i^{k-1}(x) + \frac{x_{i+k} - x}{x_{i+k} - x_{i+1}} N_{i+1}^{k-1}(x), \ 2 \le k \le n+1 \quad (4.42)$$

Note that N_i^k is a polynomial of degree $k - 1$ defined over the interval $x_i \le x \le x_{i+1}$ and gives us local control over the shape of the final B-spline curve. Further, it is easy to check that B-spline basis functions satisfy *partition of unity*, i.e., they sum up to 1 for all values of x. From Eq. (4.41), it is clear that control points p_i define the weights of different B-spline basis functions, and at any given point x, only k basis functions are non-zero. As a consequence of this, the B-spline depends locally on k nearest control points at any point x. In other words, a control point p_i influences the final curve only for the interval $x_i \le x \le t_{i+k}$.

Hence, to approximate any smooth functions as a linear combination of B-spline basis functions N_i^k, one can use least squares methods to find the unknown value of control points p_i given a sequence of B-spline interval points or knots x_i.

Example 4.3

Let us consider the problem of approximating the smooth function of Eq. (4.23) over a compact interval $[-5, 5]$ using B-splines as basis functions. As in Example 4.1, we assume that we have 5,000 pairs of observations (x, y) corrupted with random white noise of standard deviation 0.01. We assume that the approximation of $f(x)$, $\hat{f}(x)$ can be written as a linear combination of linearly independent B-spline basis functions of Eq. (4.42):

$$\hat{f}(x) = \sum_{j=0}^{n} p_j N_j^k(x) \tag{4.43}$$

To approximate the given set of measurement data, we fix a number of knots, n, to 10, and vary the order of splines from $k = 3$ to $k = 6$. We further use a least squares procedure to find unknown coefficients p_j. Fig. 4.4 shows the plot of true and approximated signals using various orders of B-spline basis functions. As expected, the approximation error decreases as we increase the order of local spline functions. Note that these results are much better than those obtained by polynomial basis functions in Chapter 1; however, the polynomial fits in Chapter 1 were fit to the entire interval, whereas the locally defined B-splines represent the curve between adjacent knots. ☐

Example 4.4

Let us consider the problem of approximating the analytical surface function of Eq. (4.26) over a compact interval $[0, 10] \times [0, 10]$ using B-splines as basis functions. As in Example 4.2, we assume that we have 10,000 pairs of observations (x_1, x_2). We assume that the approximated surface $\hat{f}(x_1, x_2)$ can be written as a linear combination of 2-D linearly independent B-spline basis functions. Further, we fix the degree of B-splines to be 6 and vary the number of knots. In particular, we consider 15×15 and 40×40 grids to generate knot vectors for the B-spline. We further use a least squares procedure to find unknown coefficients. Fig. 4.5 shows the plot of true and approximated surface and corresponding error approximation surfaces using different grid sizes for knot vectors. As expected, the approximation error decreases as we increase the the number of knots. Note that these results are better than those obtained by using wavelet basis functions as shown in Example 4.2. This is due to the fact that B-spline basis functions have a compact support, while Shannon wavelet functions do not have a compact support. ☐

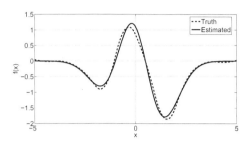

(a) 3rd Order B-Splines

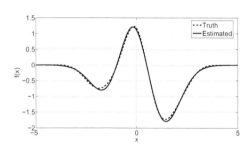

(b) 4th Order B-Splines

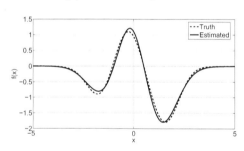

(c) 5th Order B-Splines

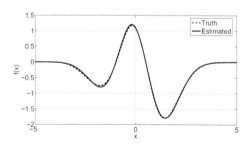

(d) 6th Order B-Splines

FIGURE 4.4
Least Squares Approximation Using B-Splines.

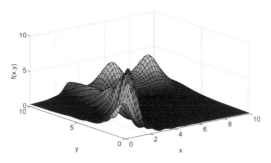

(a) Approximated Surface, 15×15 grid for knots

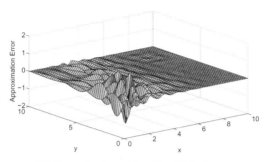

(b) Error Surface, 15×15 grid for knots

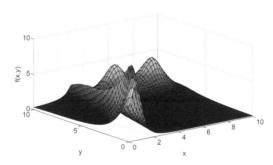

(c) Approximated Surface, 40×40 grid for knots

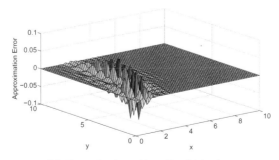

(d) Error Surface, 40×40 grid for knots

FIGURE 4.5
B-Spline Approximation of Analytical Surface Given by Eq. (4.26).

4.4   Moving Least Squares Method

Let $f(\mathbf{x}) : \mathcal{U} \subset \mathcal{R}^n \to \mathcal{R}$ be an unknown continuous function, and the moving least squares approximation, $\hat{f}(\mathbf{x})$, of the unknown function $f(\mathbf{x})$ at the point $\mathbf{x} \in \mathcal{U}$ is written as

$$\hat{f}(\mathbf{x}) = \boldsymbol{\phi}^T(\mathbf{x})\mathbf{a}(\mathbf{x}) \tag{4.44}$$

where $\boldsymbol{\phi} \in \mathcal{R}^n$ represents a complete set of basis functions, $\phi_i(\mathbf{x})$, and $\mathbf{a}(\mathbf{x}) \in \mathcal{R}^n$ is a vector of corresponding Fourier coefficients $a_i(\mathbf{x})$. Note that the amplitudes of basis functions are a function of $\mathbf{x}$ instead of being constant as in the conventional Gaussian least squares approach. To estimate the value of these unknown coefficients, a set of m distinct measurement data points $\{\mathbf{x}_i\}_{i=1,\cdots,m}$ are considered in the neighborhood $\Omega_x \subset \mathcal{U}$ of $\mathbf{x}$, and the coefficient vector $\mathbf{a}(\mathbf{x})$ is obtained by minimizing the mean square error over Ω_x.

$$J = \sum_{i=1}^{m} w_i(\mathbf{x}, \mathbf{x}_i) \left(\boldsymbol{\phi}^T(\mathbf{x}_i)\mathbf{a}(\mathbf{x}) - f_i \right)^2 \tag{4.45}$$

Here, f_i is the value of unknown function $f(\mathbf{x})$ at points $\mathbf{x}_i$. w_i is a weight function associated with the ith node such that $w_i(\mathbf{x}, \mathbf{x}_i) > 0$. In addition to positivity, the weight function w_i also satisfies the following properties:

1. The domain Ω_i of weight function w_i is a compact sub-space of $\mathcal{U}$ and $\Omega_i = \Omega_x$.

2. w_i is a monotonically decreasing function in $\|\mathbf{x} - \mathbf{x}_i\|$, i.e., as we move away from the center of weight function $\mathbf{x}_i$, the value of weight function decreases.

3. As $\|\mathbf{x} - \mathbf{x}_i\| \to 0$, $w_i \to \delta$.

Generally, a Gaussian weight function of the following form is used:

$$w_i(\mathbf{x}) = \begin{cases} \frac{e^{-(\|\mathbf{x}-\mathbf{x}_i\|/c)^2} - e^{-(r_i/c)^2}}{1 - e^{-(r_i/c)^2}}, & \|\mathbf{x} - \mathbf{x}_i\| \leq r_i \\ 0, & \|\mathbf{x} - \mathbf{x}_i\| > r_i \end{cases} \tag{4.46}$$

where r_i and c are parameters which dictate the size of domain of definition, Ω_m. As another possibility, a spline weight function is used for the MLS approximation in Ref. [53].

$$w_i(\mathbf{x}) = \begin{cases} 1 - 6\left(\frac{d_i}{r_i}\right)^2 + 8\left(\frac{d_i}{r_i}\right)^3 - 3\left(\frac{d_i}{r_i}\right)^4, & d_i = \|\mathbf{x} - \mathbf{x}_i\| \leq r_i \\ 0, & d_i > r_i \end{cases} \tag{4.47}$$

The optimum estimate $\hat{\mathbf{a}}(\mathbf{x})$ of unknown amplitude of basis functions $\mathbf{a}(\mathbf{x})$ is found by satisfying the following necessary and sufficient conditions:

Necessary Condition

$$\frac{\partial}{\partial \mathbf{a}(\mathbf{x})} J|_{\hat{\mathbf{a}}(\mathbf{x})} = \boldsymbol{\Phi}^T \mathbf{W} \mathbf{f} - (\boldsymbol{\Phi}^T \mathbf{W} \boldsymbol{\Phi}) \hat{\mathbf{a}}(\mathbf{x}) = 0 \qquad (4.48)$$

Sufficient Condition

$$\frac{\partial^2}{\partial \mathbf{a}(\mathbf{x}) \mathbf{a}(\mathbf{x}^T)} J|_{\hat{\mathbf{a}}(\mathbf{x})} = \boldsymbol{\Phi}^T \mathbf{W} \boldsymbol{\Phi} \geq 0, \ \text{(i.e., positive definite)} \quad (4.49)$$

where $\mathbf{f} \in \mathcal{R}^m$ is a vector with entries f_i, and the matrices $\boldsymbol{\Phi}$ and $\mathbf{W}$ are given by

$$\boldsymbol{\Phi}_{ij} = \phi_j(\mathbf{x}_i) \qquad (4.50)$$
$$\mathbf{W}_{ij} = w_i(\mathbf{x})\delta_{ij} \qquad (4.51)$$

Now, Eq. (4.48) yields the following solution for the optimum estimated amplitude vector $\hat{\mathbf{a}}(\mathbf{x})$ as

$$\hat{\mathbf{a}}(\mathbf{x}) = \left(\boldsymbol{\Phi}^T \mathbf{W}(\mathbf{x}) \boldsymbol{\Phi}\right)^{-1} \boldsymbol{\Phi}^T \mathbf{W}(\mathbf{x}) \mathbf{f} \qquad (4.52)$$

The necessary condition for the MLS solution to exist is that the rank of the matrix $\boldsymbol{\Phi}$ should be at least n. As a consequence of this, the support of weight function Ω_x should consist of at least n nodal points. The main difference between the conventional Gaussian least squares procedure and the moving least squares method is the choice of the weight function. In the conventional least squares procedure, the weight matrix is chosen to be the inverse of the measurement error covariance matrix and is assumed to be constant. However, in the moving least squares procedure the weight matrix is chosen such that the observations near to the evaluation point get more weightage than the observations that are far away from the evaluation point. The main drawback of the moving least squares approximation is that it is valid only at one evaluation point, $\mathbf{x}$, and therefore a new linear system dictated by Eq. (4.52) needs to be solved when the evaluation point is changed. However, if the weight functions have local support (domain), then we need to solve the linear system in the neighborhood of the evaluation point $\mathbf{x}$, thus bringing down the computational cost. Further, Levin and Wendland [54,55] have shown that if we have polynomial basis functions and the support of the weight function is local and proportional to some mesh size ρ, then we have an approximation of the order $O(\rho^d + 1)$, where d is the degree of polynomial basis functions.

Example 4.5
We once again consider the problem of approximating the smooth function of Example 4.1 over a compact interval $[-5, 5]$ using the moving least squares

method.
$$f(x) = 1.1(1 - x - 2x^2)e^{-\frac{x^2}{2}} \tag{4.53}$$

As in Example 4.1, we assume that we have 5,000 pairs of observations $(x_1, y_1), \cdots, (x_{5000}, y_{5000})$ of (x, y), corrupted with white noise of standard deviation 0.01. To study the effectiveness of the moving least squares procedure, we consider 500 evaluation points and use the weight function of Eq. (4.47) to define the support of each local approximation. We further vary the size of the approximation domain associated with each evaluation point from $r_i = 4$ to $r_i = 1$. For each local approximation, we use second order polynomial basis functions. Fig. 4.6 shows the plot of true and approximated signals using the moving least squares procedure. As expected, the approximation error decreases as we decrease the size of local approximation support. Once again, these results are much better than those obtained by polynomial basis functions in Chapter 1. This once again reinforces the advantage of multi-resolution algorithms.   ☐

Example 4.6
We again consider the problem of approximating the analytical surface of Eq. (4.26) over a compact interval, $[0, 10] \times [0, 10]$, using the moving least squares method. As in Example 4.2, we assume that we have 10,000 pairs of observations and 5,000 pairs of test data. To study the effectiveness of the moving least squares procedure, we consider a 30×30 grid of evaluation points and use the weight function of Eq. (4.47) to define the support of each local approximation. We further vary the size of the approximation domain associated with each evaluation point from $r_i = 1$ to $r_i = 0.5$. For each local approximation, we use second order polynomial basis functions. Fig. 4.7 shows the plot of true and approximated surfaces using the moving least squares procedure and corresponding approximation error surfaces. As expected, the approximation error decreases as we decrease the size of local approximation support.   ☐

4.5   Adaptive Multi-Resolution Algorithm

In this section, we present an efficient multi-resolution learning algorithm to approximate a general unknown input-output map. The main steps involved in formulating this multi-resolution learning algorithm are described as follows:

1. Given input-output data, find a simple global model which captures the global complexity at least in a coarse manner.

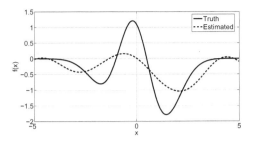

(a) $r_i = 4$

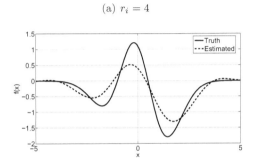

(b) $r_i = 3$

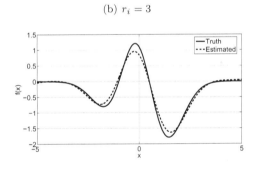

(c) $r_i = 2$

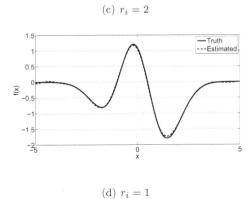

(d) $r_i = 1$

FIGURE 4.6
Moving Least Squares Approximation of Function of Eq. (4.23).

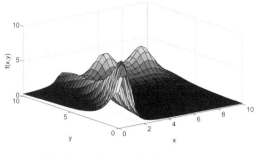

(a) Approximated Surface ($r_i = 1$)

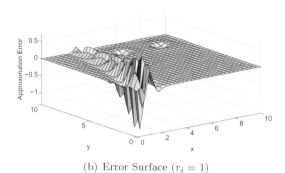

(b) Error Surface ($r_i = 1$)

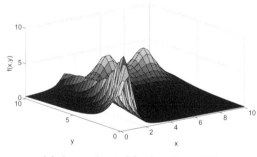

(c) Approximated Surface ($r_i = 0.5$)

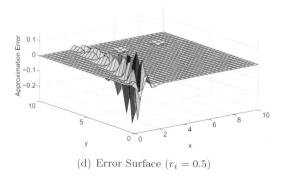

(d) Error Surface ($r_i = 0.5$)

FIGURE 4.7

Moving Least Squares Approximation of Analytical Surface Given by Eq. (4.26).

2. Refine the global model learned in the previous step until the desired approximation accuracy is achieved. To refine the global model we can introduce the local models based upon some heuristic without altering the global model.

3. To add these local models, use the "model mismatch heuristic," i.e., add local models in the region where current model errors are more than the desired accuracy.

4. Select the basis functions to describe the local models and learn their parameters using weighted statistics of local training data.

5. If the approximation errors are still large, then either change the local basis functions or introduce more local models.

This whole process is repeated until introduction of more local models does not cause any improvement in the learning of input-output mapping.

To learn the global model, we use a two-layer ANN with RBF activation functions (discussed in greater detail in Chapter 3). The main feature of the proposed learning algorithm for the RBF-based ANN is the judicious choice for locating and shaping the RBFs via a Directed Connectivity Graph approach. This approach allows a priori adaptive sizing of the network and zeroth order network pruning. In addition, it provides direction-dependent scaling and rotation of basis functions for maximal trend sensing and minimal parameter representations. Adaptation of the network parameters is done to account for online tuning, given additional measurements. To gain high resolution, the input-output data is further represented by using a family of simpler local approximations, in addition to the global RBF approximation. This is done, because the RBF approach may be defeated by the curse of dimensionality if a highly irregular, high-dimensioned system is to be approximated with high precision. The most important step in implementing the multi-resolution algorithm is to learn the local fine structure of the approximation without altering the global approximation of the input-output map. Therefore, to learn local models, we propose the use of the moving least squares algorithm; however, one is free to choose any of the methods discussed in this chapter. We introduce local models based upon the statistics of the global approximation residuals map. For regions where statistical measures of the errors (e.g., mean and standard variation) are larger than prescribed tolerances, the moving least squares process can be used to reduce approximation errors to achieve the desired resolution. To get an idea of the statistical error, the error analysis presented in the previous section can be used.

4.6   Numerical Results

To show the effectiveness of the proposed multi-resolution algorithm on a significant problem, we once again consider the problem of focal plane calibration
of a vision sensor.

4.6.1   Calibration of Vision Sensors

Vision-based sensors have found immense applications not only in the aerospace
industry, but in manufacturing inspection and assembly as well. Star tracker
cameras and vision-based sensors are primarily used to determine a spacecraft's attitude and position. In order to achieve high precision information
from these sensors, those systematic effects which tend to introduce error in
the information must be accounted for. These effects can include lens distortion and instrument aging. A lot of learning algorithms have been presented
in the literature to learn the focal plane distortion map. A detailed overview
of calibration of CCD cameras (digital cameras) can be found in Refs. [4, 5].
These papers provide a description of the various distortion mechanisms, and
review means by which these distortion mechanisms can be accounted.

The first step in the calibration process is to hypothesize an observation
model for the vision sensor. This is usually based on the physical insight regarding the particular sensor. For camera-like sensors, the following collinearity equations are used to model the projection from object space to image
space as a function of the attitude of the object:

$$x_i = -f \frac{C_{11}r_{x_i} + C_{12}r_{y_i} + C_{13}r_{z_i}}{C_{31}r_{x_i} + C_{32}r_{y_i} + C_{33}r_{z_i}} + x_0, \ i = 1, 2, \cdots, N \qquad (4.54)$$

$$y_i = -f \frac{C_{21}r_{x_i} + C_{22}r_{y_i} + C_{23}r_{z_i}}{C_{31}r_{x_i} + C_{32}r_{y_i} + C_{33}r_{z_i}} + y_0, \ i = 1, 2, \cdots, N \qquad (4.55)$$

where C_{ij} are the unknown elements of attitude matrix $\mathbf{C}$ associated to the
orientation of the image plane with respect to some reference plane, f is the
known focal length, (x_i, y_i) are the known image space measurements for the
i^{th} line of sight, $(r_{x_i}, r_{y_i}, r_{z_i})$ are the known object space direction components
of the ith line of sight and N is the total number of measurements. x_0 and
y_0 refer to the principal point offset. Generally, the focal plane calibration
process is divided into two major parts:

1. Calibration of principal point offset (x_0, y_0) and focal length (f).

2. Calibration of the non-ideal focal plane image distortions due to all other
   effects (lens distortions, misalignment, detector alignment, etc.).

The implicit pinhole camera model is not exact so we need to find the best
effective estimates of principal point offset (x_0, y_0) and focal length (f). However, the principal point offset is obviously correlated with the inertial pointing

of the boresight. In our earlier work [6, 56], we proposed the "attitude independent" approach (essentially, based upon interstar angle measurements) to eliminate this difficulty. While this approach leads to reduced observability of (x_0, y_0), we find redundant measurement are sufficient to determine good estimates for (x_0, y_0) and f. Additionally, we need one attitude independent algorithm to identify the objects in the image plane. In Ref. [57] we presented a non-dimensional star identification algorithm for the spacecraft attitude determination problem using a star camera to identify the stars without any attitude knowledge. For any focal plane calibration algorithm to work, the uncalibrated sensor's errors must be sufficiently small so that the non-dimensional star identification algorithm works reliably. After the first calibration is achieved, our studies indicate that any of the several star identification algorithms work reliably. While the "how to get started" issue is important, we choose not to add to this discussion in this chapter, and implicitly assume that the Eqs. (4.54) and (4.55) are sufficiently precise with the initial estimates of (x_0, y_0) and f. In this chapter, we only demonstrate the application of the multi-resolution approximation procedure discussed in the previous section to learn higher order image distortion effects.

4.6.2 Simulation and Results

To demonstrate the effectiveness of the multi-resolution learning algorithm, an $8° \times 8°$ Field of View (FOV) star camera is simulated by using the pinhole camera model dictated by Eqs. (4.54) and (4.55) with principal point offset of $x_0 = 0.75$ mm and $y_0 = 0.25$ mm. The focal length of the star camera is assumed to be 64.2964 mm.

For simulation purposes, the spacecraft is assumed to be in a low Earth orbit tumbling with the following angular velocity about the sensor axis aligned to the z-axis of the spacecraft body frame.

$$\omega = \{ \omega_0 sin(\omega_0 t) \ \omega_0 cos(\omega_0 t) \ \omega_0 \}, \omega_0 = 10^{-3} rad/sec \qquad (4.56)$$

Assuming the star camera imaging frequency to be 1 Hz, the star data is generated for 2 hr motion of the spacecraft.

The true lens distortion is assumed to be given by the following models [5]:

$$\mathbf{\Phi} = \{ r \ r^2 \ r^3 \ r^4 \ \}; \ \delta x = x \mathbf{\Phi}^T \mathbf{a} \ \& \ \delta y = y \mathbf{\Phi}^T \mathbf{b} \qquad (4.57)$$

where
$$r = \sqrt{x^2 + y^2} \qquad (4.58)$$

To learn the distortion map, we need some measure of the measurement error that can be used to model the systematic focal plane distortion map. Further, this model of the distortion map can be used to correct measurements. The best estimate of attitude and cataloged vectors are "run through" Eqs. (4.54) and (4.55) to predict δx_i and δy_i, the differences from measurements. Initially, the attitude, C_{ij}, in Eqs. (4.54) and (4.55) was perturbed (not only by

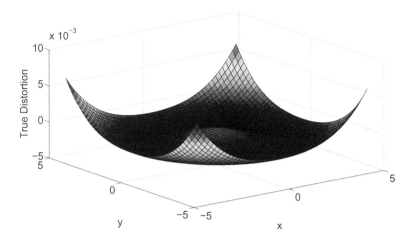

FIGURE 4.8
True Distortion Map.

random measurement errors, but also by calibration errors) because no distortion calibration has been applied on the first pass. We mention that after
the first approximate calibration, δx and δy estimated distortions should be
added to correct the measured x_i, y_i in Eqs. (4.54) and (4.55) before forming line-of-sight vectors that are used to estimate the attitude matrix $\mathbf{C}$. In
Ref. [56], we have shown how the second order calibration perturbations of
$\mathbf{C}$ get reduced as the δx, δy are refined. We mention that the reason for the
convergence of the calibration process is that the moderate-sized calibration
errors perturb the "rigid body" attitude estimate, but the residual errors in
measurements minus the prediction of Eqs. (4.54) and (4.55) still have most
of the high order distortion effects. For simulation purposes, the net attitude
error due to residual calibration and sensor noise is sought to be 10 μrad or
smaller.

Fig. 4.8 shows the surface plot of the true distortion map given by Eq. (4.57)
with the following value of $\mathbf{a}$

$$\mathbf{a} = \left\{ 5 \times 10^{-4} \ -5 \times 10^{-4} \ 8 \times 10^{-4} \ -8 \times 10^{-4} \right\}^T \qquad (4.59)$$

From this figure, it is clear that the distortion surface amounts to calibration
errors of the order of 10^{-3} radians. We seek to reduce these errors two orders
of magnitude to the order of 10 μrad by using the multi-resolution algorithm
discussed in the previous section. In the next section we present the global
approximation result using the DCG algorithm as discussed in Chapter 3,
followed by local approximation (based upon the "model mismatch heuristic")
results using the MLS algorithm.

4.6.3 DCG Approximation Result

In this section, we present the global approximation result for the distortion map shown in Fig. 4.8 using the DCG algorithm as discussed in Chapter 3. To approximate the distortion map given by Eq. (4.57), the input region is divided into a total of 4 square regions (2 in each direction). Then according to the procedure listed in Chapter 3, we generate a directed connectivity graph of the local maxima and minima in each sub-region that finally lead to 8 radial basis functions that have approximation errors of the order of $O(10^{-3})$.

Fig. 4.9 shows the approximation error for the training set. From this figure, it is clear that the DCG-learned RBF network is able to approximate the distortion map with very good accuracy using only 8 radial basis functions. The DCG algorithm is tested upon uniformly distributed points in the focal plane. Fig. 4.10(a) shows the approximated distortion map learned by the DCG algorithm, whereas Fig. 4.10(b) shows the approximation error surface. From these results, it is clear that although the DCG approach has done a good job in learning the shape of the distortion map and reducing the errors by about one order of magnitude, the approximated map still has some large amplitude errors.

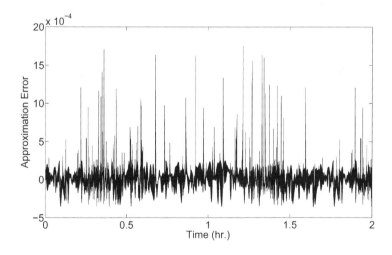

FIGURE 4.9
Global Approximation Results Using the DCG Algorithm for the Training Data Set.

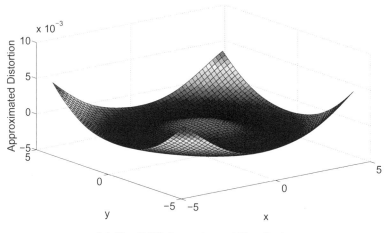

(a) The DCG Approximated Test Surface

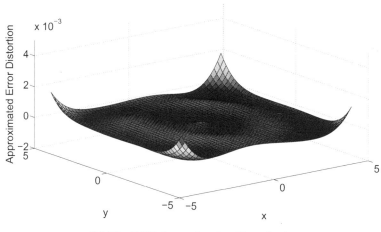

(b) The DCG Approximation Error Surface

FIGURE 4.10
Global Approximation Results Using the DCG Algorithm for the Test Data
Set.

4.6.4   Local Approximation Results

In the previous section we presented the global approximation results for the distortion map given by Eq. (4.57). In this section, we present the results which show how the multi-resolution-based learning algorithm improves the global approximation.

From the results presented in the previous section, it is clear that with only a global DCG-based RBFN approximation we are able to learn the distortion map with an accuracy smaller than $O(10^{-3})$. To achieve the desired accuracy of the order of $O(10^{-5})$, we invoke the multi-resolution MLS algorithm to correct the approximation error surfaces from the DCG algorithm.

Fig. 4.11(a) shows the DCG approximation error surface corrected by the MLS algorithm, whereas Fig. 4.11(b) shows the net approximation error surfaces. From these figures, it is clear that with the help of local approximation using the MLS algorithm, we can further reduce the global approximation errors by two orders of magnitude.

From these results, we can conclude that the multi-resolution-based local approximation helps us have a flexible and adaptive calibration process that does not rely on simply guessing with a distortion map.

4.7   Summary

In this chapter we introduced conventional multi-resolution algorithms and briefly discussed their use in function approximation. The multi-resolution properties of these algorithms have led to broadly useful approximation approaches that have good local approximation properties for any given input-output data. Although global-local separation of concern leads to immensely improved approximation algorithms, for a specific multi-resolution algorithm one cannot use different basis functions to obtain different local approximations without introducing discontinuity across the boundary of different local regions. For example, in the case of wavelet-based approximation, one should use the same wavelet function at different resolution levels, and similarly, in the case of B-spline, one is restricted to using only polynomial basis functions in various intervals. Hence, there is a need for rigorous methods to merge different independent local approximations to obtain a desired order, globally continuous approximation. Finally, an efficient adaptive learning algorithm is developed which not only has the global approximation capability of the ANN, but also has the multi-resolution capability. Computational experiments are conducted to evaluate the utility of the developed multi-resolution algorithm, and simulation results provide compelling evidence and a basis for optimism.

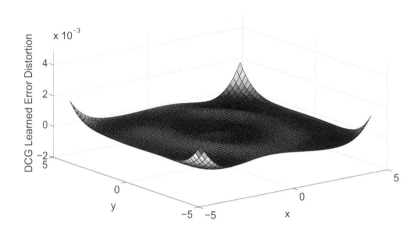

(a) The DCG Approximation Error Surface Learned by the MLS Algorithm

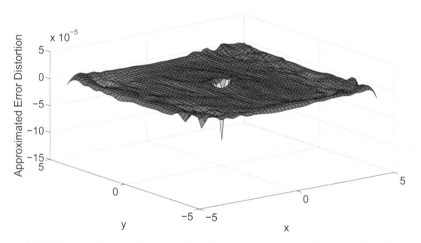

(b) Net Approximation Error Surface Using the DCG and the MLS Algorithms

FIGURE 4.11
Multi-Resolution Approximation Results.

5

Global Local Orthogonal Polynomial MAPping (GLO-MAP) in N Dimensions

Facts are many, but the truth is one.

Rabindranath Tagore

5.1 Introduction

Multi-resolution is an important property for any approximation algorithm. Whether one is compressing satellite images, trying to solve Partial Differential Equations (PDEs), or modeling some irregular function, there is broad interest in multi-resolution approximation. Spline approximations, wavelets and Finite Element Methods (FEM), are the most commonly used multi-resolution algorithms. Qualitatively, multi-resolution refers to the ability of an approximation to represent macroscopic features as well as very fine structure features. The general strategy of all multi-resolution algorithms is similar; however, due to some additional properties specific to each algorithm, some algorithms are more suited for specific applications. For example, wavelets-based approximation methods are well suited for image processing and FEM are generally accepted as being effective for solving PDEs. These approximation methodologies are also shown to be compatible with a wide variety of disciplines, such as continuous function approximation, dynamic system modeling, time series prediction and image processing. The multi-resolution properties of these algorithms have led to broadly useful approximation approaches that have good local approximation properties for any given input-output data. Multi-resolution approximation enables separation of concern between model complexity, accuracy and compression ratio. It also enables accurate characterization of noise and uncertainty in the data, leading to robust approximation models. The basis functions to construct local approximations at various granularity levels are usually chosen from a space of polynomial functions, trigonometric functions, splines, or radial basis functions. Ideally, one may prefer to choose these basis functions, based on prior knowledge about the problem, or based solely upon local approximability.

For example, if the local behavior is known to be highly oscillatory, harmonic functions with frequencies tuned to capture the actual system oscillations can be used in the basis set. Alternatively, these local models can also be constructed intelligently by studying local behavior of the input-output map by the use of methods like Principal Component Analysis or Fourier Decomposition [58–60]. Such special functions may be introduced either by themselves, or to supplement a previously existing approximation. While such freedom provides great flexibility and can immensely improve the approximability, it generally prevents the basis functions from constituting a conforming space; i.e., the inter-element continuity of the approximation is not ensured. Following traditional approaches, the specific multi-resolution approximation basis functions used to obtain different local approximations cannot be independent of each other without introducing discontinuity across the boundary of contiguous local regions. Hence, there is a need for rigorous methods to merge different independent local approximations to obtain a desired order, globally continuous, approximation.

In this chapter we introduce an important modeling algorithm that can be utilized for a large number of different engineering applications, while considering the above-mentioned disadvantages of the best existing methods. A key motivation underlying these developments is to establish a more general, rigorous, and computationally attractive way to construct a family of local approximations. The main idea discussed is a weighting function technique [61,62] that generates a global family of overlapping preliminary approximations whose centroids of validity lie on at the vertices of an N-dimensional pseudo-grid. These natural local models are defined *independently of each other* by the use of classical basis functions (like RBF, Fourier series, polynomials, wavelets etc.), and any a priori information that we may have about local characteristics of the given input-output data. These sub-domains, where the preliminary approximations are valid, are constructed such that they overlap and the overlapping approximations are averaged over the overlapped volume to determine final local approximations. A novel averaging method is introduced that ensures these final approximations are globally piecewise continuous, with adjacent approximations determined in an analogous averaging process, to some prescribed order of partial differentiation. The continuity conditions are enforced by using a unique set of weighting functions in the averaging process, *without constraining the preliminary approximations being averaged*. The weight functions are designed to guarantee the global continuity conditions while retaining *near complete freedom on the selection of the generating local approximations*. As will be evident, the new method generalizes all of the existing MRA approaches, since any of these can be used to establish the preliminary local approximations.

The chapter is organized as follows. First, we introduce and illustrate the basic ideas underlying the proposed algorithm, followed by systematic development of the algorithm and theoretical justification that proves the probabilistic truth that the new averaging process is unbiased, and the covariance

of the averaged approximations is smaller than the generating approximations. Finally, the proposed approach is validated by considering different
engineering applications.

5.2   Basic Ideas

To motivate the results in this chapter, consider Fig. 5.1.* Here we have
64,000 noisy measurements of an irregular function, $F(x, y)$. These happen
to be stereo ray intersection measurements from correlation of stereo images
of topography near Fort Sill, Oklahoma [1]; however, they could be measurements of any complicated, irregular function for which a single global algebraic
expression would likely be intractable. Suppose that it is desired to obtain a
smooth, least squares approximation of this function, perhaps with additional
constraints imposed (e.g., in this case, the stereo correlation measurement
process fails reliably over water, so the large spurious noise spikes over lakes
Latonka and Elmer Thomas, where reliable stereo correlation is not possible, but can be replaced by a constraint that the lake surface be a known
elevation). In lieu of a single global and necessarily complicated function, it
is desired to represent the function using a family of simpler local approximations. Such local approximations would be a much more attractive basis
for local analysis. Alternatively, one may think of the local approximations
as Taylor series approximations (each evaluated at a local expansion point
on a grid), or as any local approximations obtained from local measurements.
However, if the local approximations are introduced without taking particular
care, they will almost certainly disagree in the value estimated for $F(x, y)$ and
the derivatives thereof at any arbitrary point, although the discrepancies may
be small. In other words, global continuity is not assured unless we introduce
methodology to guarantee the desired continuity properties. These challenges
are compounded in higher dimensions if the usual local approximation approaches are used. It is desired to determine a piecewise continuous global
family of local least squares approximations, while having the freedom to vary
the nature (e.g., mathematical basis functions and degrees of freedom) of the
local approximations to reflect possibly large variations in the roughness of
$F(x, y)$. While we are introducing the ideas in the setting of a data-fitting
problem in a 2-dimensional space, the results are shown later in this book
to be of much broader utility, and to generalize fully to approximation in an
N-dimensional space, including opening a door to a flexible new method for
solving high dimensional partial differential equations.

---

*Please see the Appendix for the color version of this figure.

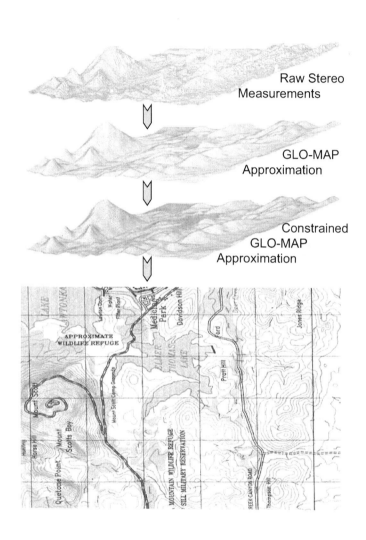

FIGURE 5.1
Approximation of Irregular Functions in 2 Dimensions.

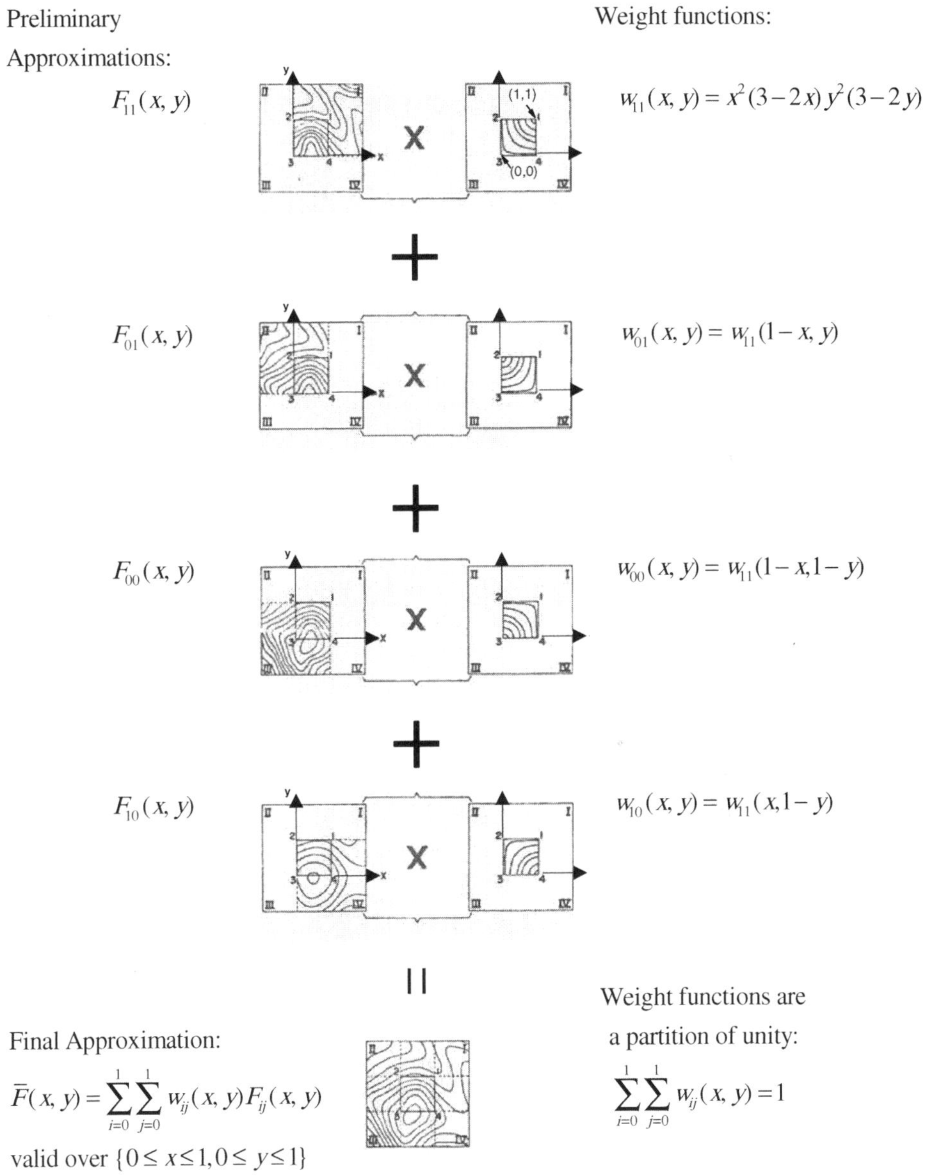

Preliminary
Approximations:

Weight functions:

$F_{11}(x, y)$

$w_{11}(x, y) = x^2(3-2x)\,y^2(3-2y)$

+

$F_{01}(x, y)$

$w_{01}(x, y) = w_{11}(1-x, y)$

+

$F_{00}(x, y)$

$w_{00}(x, y) = w_{11}(1-x, 1-y)$

+

$F_{10}(x, y)$

$w_{10}(x, y) = w_{11}(x, 1-y)$

||

Final Approximation:

Weight functions are
a partition of unity:

$$\bar{F}(x, y) = \sum_{i=0}^{1}\sum_{j=0}^{1} w_{ij}(x, y)F_{ij}(x, y)$$

$$\sum_{i=0}^{1}\sum_{j=0}^{1} w_{ij}(x, y) = 1$$

valid over $\{0 \le x \le 1, 0 \le y \le 1\}$

FIGURE 5.2
Qualitative Representation of the Averaging Process in 2 Dimensions.

With reference to Fig. 5.2, we summarize some features of the weighting function approach to approximation in two dimensions. We prove these qualitative statements later in this chapter. Using Fig. 5.2, we introduce several qualitative observations. Note the attractive properties of the weight functions: At any of the contiguous four vertices, we see the weight function (associated with the function whose centroid of validity is a given vertex) is unity, while the other three weight functions are zero at that vertex. Note further that the weight functions have a qualitative bell shape, but fair into a square base, the zero contour being the boundary opposite (e.g., 2-3-4) to the vertex (e.g., point 1) where the weight has a unit value. We will show that the four overlapping weight functions constitute a *partition of unity*; they add to unity everywhere in the overlapping unit region (which guarantees an unbiased approximation). Furthermore, note that along any boundary, only the two weight functions associated with the two approximations centered at the end points of that boundary are non-zero along that boundary, while the other two weight functions are zero (the partial derivatives of the other two weight functions are also along these boundaries). These continuity arguments on the averaged approximation of the function can be extended readily to corresponding properties on their partial derivatives: The averaged approximation osculates in value and partial derivatives with the four preliminary approximations at their corresponding vertices, and the function and both partial derivatives along any boundary are a weighted average of the corresponding two functions associated with the end point of that boundary. Collectively, these observations lead to rigorous piecewise continuity of the averaged approximations, while leaving the user free to choose any preliminary local approximations desired or needed. These qualitative observations will be developed systematically in the subsequent sections and extended rigorously to approximation with arbitrary order continuity in an N-dimensional space.

5.3   Approximation in 1, 2 and N Dimensions Using Weighting Functions

The essential ideas can be introduced rigorously in a 1-dimensional piecewise approximation problem. The notations are developed for the 1-dimensional problem such that the generalization is most straightforward. With reference to Fig. 5.3,* we discuss the 1-dimensional problem. An arbitrary set of knots (vertices), $\{{}^1X, {}^2X, \cdots, {}^KX, \cdots\}$, are introduced at a uniform distance h apart; a non-dimensionalization of x is introduced as a local coordinate

---

*Please see the Appendix for the color version of this figure.

$-1 \leq {}^I x \overset{\Delta}{=} (X - {}^I X)/h \leq 1$, centered on the Ith vertex $X = {}^I X$. The local weighted average approximation is introduced as

$$\bar{F}_I(X) = w({}^I x)F_I(X) + w({}^{I+1} x)F_{I+1}(X), \text{ for } 0 \leq^I x < 1 \qquad (5.1)$$

where the weighting functions $w(x)$ used to average (blend) the two adjacent preliminary local approximations $\{F_I(X), F_{I+1}(X)\}$ are as yet unspecified. We prefer that the preliminary approximations $\{F_1(X), \cdots, F_K(X), \cdots\}$ *be left completely arbitrary*, so long as they are smooth and represent the local behavior of $F(X)$ well. As developed in Ref. [63], the weight function can be selected to guarantee that the averaged approximation $\bar{F}(X)$ osculates with $F_I(X)$ in value and first derivative as $X \rightarrow {}^I X$, and likewise $\bar{F}(X)$ osculates with $F_{I+1}(X)$ in value and first derivative as $X \rightarrow {}^{I+1} X$. Note that the shifted weight functions add to unity, as they must for an unbiased estimate, e.g., $w({}^I x) + w({}^I x - 1) = 1$, or $w({}^I x - 1) = 1 - w({}^I x)$. Observe that ${}^{I+1} x = {}^I x - 1$, so if $0 \leq {}^I x \leq 1$, $-1 \leq {}^{I+1} x = {}^I x - 1 \leq 0$. Note also the first derivative of the average of Eq. (5.1) at an arbitrary point is

$$\frac{d\bar{F}_I(X)}{dx} = w({}^I x)\frac{dF_I(X)}{dx} + w({}^{I+1} x)\frac{dF_{I+1}(X)}{dx} + \frac{dw({}^I x)}{dx}F_I(X)$$
$$+ \frac{dw({}^{I+1} x)}{dx}F_{I+1}(X) \qquad (5.2)$$

Thus the requirement that the weighted average approximation of Eq. (5.1) form a continuous globally valid model leads to the following boundary conditions on yet-to-be-defined weighting functions:

$$\text{at } x = 0 : \begin{cases} w(0) = 1 \\ \frac{dw(x)}{dx}\Big|_{x=0} = 0 \end{cases}, \quad \text{at } x = 1 : \begin{cases} w(1) = 0 \\ \frac{dw(x)}{dx}\Big|_{x=1} = 0 \end{cases} \qquad (5.3)$$

With these boundary conditions, the first term of Eq. (5.1) reduces to $F_I(X)$ as ${}^I x \rightarrow 0$, and likewise, only the first term of Eq. (5.2) contributes as ${}^I x \rightarrow 0$. Analogous osculation arguments hold at the right end of the interval. In general, the requirement that the weighted average approximation in Eq. (5.1) form an mth-order continuous globally valid model and the additional requirement of unbiased approximation leads to the following boundary value problem that uniquely defines the necessary weighting functions:

1. The first derivative of the weighting function must have an mth-order osculation with $w(0) = 1$ at the centroid of its respective local approximation.

$$w(0) = 1$$
$$\frac{d^k w}{dx^k}\Big|_{x=0} = 0 \quad k = 0, 1, \cdots, m \qquad (5.4)$$

2. The weighting function must have an $(m+1)^{th}$-order zero at the centroid of its neighboring local approximation.

$$w(1) = 0$$
$$\frac{d^k w}{dx^k}\Big|_{x=1} = 0 \quad k = 0, 1, \cdots, m \qquad (5.5)$$

3. The sum of two neighboring weighting functions must be unity over the entire closed interval between their corresponding adjacent local functional approximations.

$$w(^Ix) + w(^Ix - 1) = 1 \quad \forall \ x, \ -1 \leq x \leq 1 \tag{5.6}$$

It should be noted that the first two boundary conditions are sufficient to ensure that the global function reduces exactly to the local approximations at their centroids, not only in their value but also in their first m partial derivatives.

A family of weight functions satisfying the aforementioned boundary value problem can be obtained by assuming the following particular form for the weighting function,

$$w(x) = 1 - J(x) \tag{5.7}$$

where $J(x)$ is a polynomial in the independent variable whose first derivative is given by the following expression:

$$\frac{dJ(x)}{dx} = Cx^m(1 - x)^m \tag{5.8}$$

It should be noted that this particular form for the weighting function is in accordance with the fact that first m partial derivatives of the weighting function vanish at end points and $w(0) = 1$. Now the remaining boundary conditions on the weighting function, $w(x)$, will completely define the constants in Eq. (5.8).

$$J(1) = C \int_0^1 x^m(1 - x)^m dx = 1 \tag{5.9}$$

That means the appropriate value for the constant, C, is given by

$$C = \left[\int_0^1 x^m(1 - x)^m dx \right]^{-1} \tag{5.10}$$

Now, using the fact that integral expression on the RHS is a Eulerian integral of the first kind [64], the constant C is given by the following expression:

$$C = \frac{(2m + 1)!}{(m!)^2} \tag{5.11}$$

The general form for the weighting function can now be written as

$$w(x) = 1 - \frac{(2m + 1)!}{(m!)^2} \int_0^x x^m(1 - x)^m dx \tag{5.12}$$

Further, the binomial theorem allows us to expand the above integrand.

$$w(x) = 1 - \frac{(2m+1)!}{(m!)^2} \int_0^x \sum_{n=0}^m x^m x^{m-n}(-1)^n dx \qquad (5.13)$$

Now, integrating the above expression term by term yields the following expression for the weighting function, $w(x)$:

$$w(x) = 1 - K \sum_{n=0}^m A_n x^{2m-n+1} \qquad (5.14)$$

where K and A_n are given by the following expressions:

$$K = \frac{(2m+1)!(-1)^m}{(m!)^2}, \quad A_n = \frac{(-1)^n \, {}^m C_n}{2m-n+1} \qquad (5.15)$$

Finally, to obtain the expression for the weighting function in the interval $[-1,1]$ instead of $[0,1]$, the absolute value of x is used as the independent variable rather than x. The lowest order weight function (for $m = 1$) can be shown to be simply

$$w(x) = \left\{ \begin{array}{l} 1 - x^2(3+2x), \ -1 \le x < 0 \\ 1 - x^2(3-2x), \ 0 \le x \le 1 \end{array} \right\} = 1 - x^2(3 - 2|x|) \qquad (5.16)$$

These are the functions plotted in Fig. 5.3. It should be noted that the weight function given by Eq. (5.16) is a non-negative continuous function defined on a locally compact subset of approximation space. To be more precise, the weight functions obtained by solving the boundary value problem have the following properties, which enable an important meshless approximation and interpolation algorithm.

1. The domain of the weight function w is a compact space.

2. $w(x) > 0, \ \forall \ x \in (-1, \ 1)$.

3. $w(x) = 0, \ |x| \ge 1$.

4. $w(x)$ is a monotonically decreasing function of x, $\forall \ x \in (-1, \ 1)$.

In the event that discrete measurements of $F(X)$ are available, the preliminary approximations $\{F_1(X), F_2(X), \cdots, F_K(X), \cdots\}$ are fit to data subsets in the $\Delta X = \pm h$ regions centered on $\{ {}^1X \ {}^2X \cdots {}^KX \cdots \}$. It is evident that the final approximation on each interval is the average of overlapping weighted least squares approximations, fit to shifted data lying within $\pm h$ of the vertices. For equally precise measurements of $F(X)$, the least squares process should use the same weight functions of Eq. (5.3). If the measurements are made with unequal expected precision, then the statistically justified weights

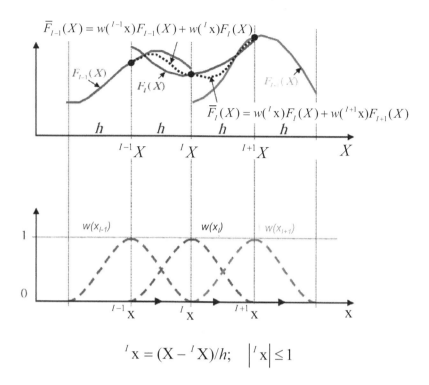

FIGURE 5.3
Weighting Function Approximation of a 1-Dimensional Function.

should be scaled using the weights of Eq. (5.16). Note the qualitative justification: "If one least squares fit is good, an unbiased average of two should
be better." Later in this chapter, we prove the probabilistic veracity of this
qualitative observation. Observe that simply through choosing the judicious
weight functions of Eq. (5.16) we are guaranteed global piecewise continuity for
all possible continuous local approximations $\{F_1(X), F_2(X), \cdots, F_K(X), \cdots\}$.
One retains the freedom to vary the degree and/or the mathematical form of
the local approximations, as needed, to fit the local behavior of $F(X)$ and
rely upon the weight functions to enforce continuity.

A most important characteristic of this weighting function averaging process is that it generalizes fully to N dimensions without as severe a "curse
of dimensionality" that accompanies generalizations of virtually all known
analysis methods to higher dimensions. The generalization to 2 dimensions is
amazingly straightforward. Introduce notation for the local approximations
$\{F_{11}(X_1, X_2), \cdots, F_{I_1 I_2}(X_1, X_2), \cdots\}$ constructed such that they are valid

over $(2h) \times (2h)$ regions centered on the vertices $\{(\,^1X_1,\,^1X_2), (\,^1X_1,\,^2X_2), \cdots, (\,^{I_1}X_1,\,^{I_2}X_2), \cdots\}$. Given four contiguous vertices,

$$
\begin{array}{ll}
(\,^{I_1}X_1,\,^{I_2+1}X_2 = \,^{I_2}X_2 + h) & (\,^{I_1+1}X_1 = \,^{I_1}X_1 + h,\,^{I_2+1}X_2 = \,^{I_2}X_2 + h) \\
(\,^{I_1}X_1,\,^{I_2}X_2) & (\,^{I_1+1}X_1 = \,^{I_1}X_1 + h,\,^{I_2}X_2)
\end{array}
$$

$$(5.17)$$

The corresponding four preliminary approximations valid in the $(2h) \times (2h)$ regions centered at the contiguous four nodes are denoted as:

$$
\begin{array}{ll}
F_{I_1,I_2+1}(X_1, X_2) & F_{I_1+1,I_2+1}(X_1, X_2) \\
F_{I_1,I_2}(X_1, X_2) & F_{I_1+1,I_2}(X_1, X_2)
\end{array}
$$

$$(5.18)$$

The final averaged approximation valid within the $h \times h$ region bounded by the four vertices of Eq. (5.17) is given by

$$
\bar{F}_{I_1,I_2}(X_1, X_2) = \sum_{i_1=0}^{1} \sum_{i_2=0}^{1} w_{i_1,i_2}(\,^{I_1+i_1}x_1,\,^{I_2+i_2}x_2) F_{I_1+i_1,I_2+i_2}(X_1, X_2)
$$

$$(5.19)$$

where it can be verified that choosing the weight functions as the product of 1-dimensional weight functions as

$$
w_{i_1,i_2}(\,^{I_1+i_1}x_1,\,^{I_2+i_2}x_2) = w(\,^{I_1+i_1}x_1)w(\,^{I_2+i_2}x_2) \tag{5.20}
$$

This means that these functions are a *partition of unity* that satisfy

$$
\sum_{i_1=0}^{1} \sum_{i_2=0}^{1} w_{i_1 i_2}(\,^{I_1+i_1}x_1,\,^{I_2+i_2}x_N) = 1 \tag{5.21}
$$

to give an unbiased average. It can be verified that $w_{i_1 i_2}(\,^{I_1+i_1}x_1,\,^{I_2+i_2}x_N)$ satisfies the following properties:

$$
w_{i_1 i_2}(0,0) = 1
$$

$$
\frac{\partial w_{i_1 i_2}}{\partial x_j}\bigg|_{0,0} = 0, \ j = 1, 2
$$

$$
w_{i_1 i_2}(0, x_2) = 0, \ -1 \le x_2 \le 1
$$

$$
w_{i_1 i_2}(x_1, 0) = 0, \ -1 \le x_1 \le 1
$$

$$
\frac{\partial w_{i_1 i_2}}{\partial x_j}\bigg|_{0,x_2} = 0, \ -1 \le x_2 \le 1, \ j = 1, 2
$$

$$
\frac{\partial w_{i_1 i_2}}{\partial x_j}\bigg|_{x_1,0} = 0, \ -1 \le x_1 \le 1, \ j = 1, 2
$$

These boundary conditions, along with Eq. (5.21), are sufficient to ensure that the weighted average of Eq. (5.19) leads to global piecewise continuity and an unbiased estimator [62,65]. Note that if we use a common origin (the

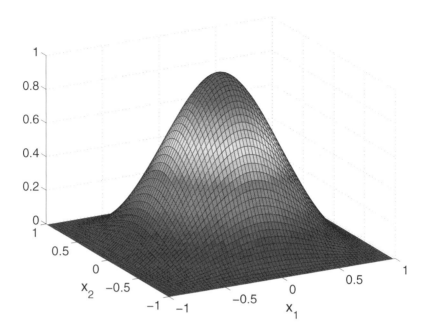

FIGURE 5.4
Weighting Function $w_{0,0}(x_1, x_2) = w(x_1)w(x_2)$ for 2-Dimensional Approximation.

lower left vertex) for all four weight functions, then the one centered on the origin (for $m = 1$) is

$$w_{0,0}(x_1, x_2) = [1 - x_1^2(3 \mp 2x_1)][1 - x_2^2(3 \mp 2x_2)]$$
$$\text{the minus (plus) sign is for } x_i > 0 \ (x_i < 0) \text{ or}$$
$$\equiv [1 - x_1^2(3 - 2|x_1|)][1 - x_2^2(3 - 2|x_2|)] \tag{5.23}$$

The remaining three weight functions are simply obtained by translating this function to the other three vertices as

$$w_{1,0}(x_1, x_2) = w_{0,0}(x_1 - 1, x_2)$$
$$w_{0,1}(x_1, x_2) = w_{0,0}(x_1, x_2 - 1) \tag{5.24}$$
$$w_{1,1}(x_1, x_2) = w_{0,0}(x_1 - 1, x_2 - 1)$$

These four overlapping weight functions are shown in Fig. 5.5. The central unit square of Fig. 5.5 is the focus of this figure. It is the region in which the final averaged approximation of Eq. (5.19) is valid. The process can be shifted by one unit cell in any direction and continuity arguments will lead to the conclusion that the adjacent final averaged approximations match both in value and partial derivatives along their common boundaries.

We see the weight function of Fig. 5.2 from Refs. [63,66] is obtained to within the obvious notation changes. The reason for adopting the above notations is that the generalization to N dimensions follows easily from the above pattern. The N-dimensional generalizations of Eqs. (5.19) and (5.20) are

$$\bar{F}_{I_1,\cdots,I_N}(X_1,\cdots,X_N) = \sum_{i_1=0}^{1}\sum_{i_2=0}^{1}\cdots\sum_{i_N=0}^{1}\left(w_{i_1,\cdots,i_N}(^{I_1+i_1}x_1,\cdots,^{I_N+i_N}x_N)\right.$$
$$\left. F_{I_1+i_1,\cdots,I_N+i_N}(X_1,\cdots,X_N)\right) \tag{5.25}$$

and

$$w_{i_1,i_2,\cdots,i_N}(^{I_1+i_1}x_1,\cdots,^{I_N+i_N}x_N) = \prod_{i=1}^{N}w(^{I_i+i_i}x_i) \tag{5.26}$$

$$\sum_{i_1=0}^{1}\sum_{i_2=0}^{1}\cdots\sum_{i_N=0}^{1}w_{i_1,\cdots,i_N}(^{I_1+i_1}x_1,\cdots,^{I_N+i_N}x_N) = 1 \tag{5.27}$$

The partition of unity constraint of Eq. (5.27) is required for an unbiased average in Eq. (5.25). Further, it can be verified that the unbiased average requirement of Eq. (5.27) is satisfied everywhere in the hypercube where averaged final approximation $\bar{F}_{I_1\cdots,I_N}(X_1,\cdots,X_N)$ of Eq. (5.25) is valid.

The above approximation approach, and minor variations of it, has been used in a wide variety of modeling problems, including mathematical modeling of topography, the earth's gravity field, the focal plane distortions of star cameras, modeling the input-output behavior of synthetic jet actuators, and many other problems in approximation theory, geophysics, engineering, and applied science (see Refs. [56, 61, 63, 66–70]). We note that it is relatively straightforward to accommodate non-uniform meshes, but this case is not addressed in the present chapter to avoid notional complexity. While the weighting function approach has much in common with finite element methods, note the distinction: Whereas conventional FEM methods interpolate nodal values of some distributed quantity into the continuous domain of the finite elements, this weighting function approach instead *averages overlapping local approximations in such a way that piecewise continuity is achieved*, with the user free to choose the local approximations. The degree of the local approximations can be adaptively modified to enhance convergence. Finally, it must be pointed out that one major drawback of the conventional FEM-based approach is the generation of a mesh for higher dimensional spaces. However, the use of specially designed weighting functions results in a meshless technique to alleviate some of the problems related to generating meshes for high dimensioned systems.

The weight functions given above (e.g., Eqs. (5.16), (6.13)) guarantee first order continuity. The generalized weight functions that guarantee arbitrary

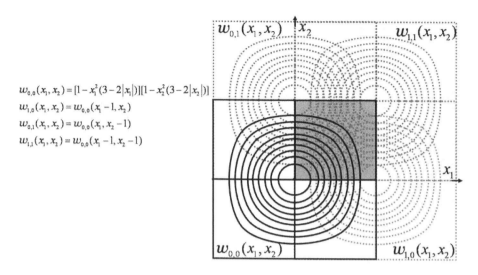

$w_{0,0}(x_1, x_2) = [1 - x_1^2(3 - 2|x_1|)][1 - x_2^2(3 - 2|x_2|)]$
$w_{1,0}(x_1, x_2) = w_{0,0}(x_1 - 1, x_2)$
$w_{0,1}(x_1, x_2) = w_{0,0}(x_1, x_2 - 1)$
$w_{1,1}(x_1, x_2) = w_{0,0}(x_1 - 1, x_2 - 1)$

FIGURE 5.5

Four Contiguous, Overlapping Weighting Functions for 2-Dimensional Approximation.

order continuity are given in Table 5.1. Only the weight function centered at the origin is tabulated, the other $2^N - 1$ weight functions are obtained by simply shifting the function using the origin translations to the other $2^N - 1$ vertices of the hypercube, analogous to Eq. (5.24), e.g., using the $2^N - 1$ origin translations,

$$\{(0, 0, 0, \ldots, 0, 0, 1), (0, 0, 0, \ldots, 0, 1, 0), \ldots (1, 1, 1, \ldots, 1, 1, 1)\} \qquad (5.28)$$

The weight functions for the first three orders of continuity, for 1- and 2-dimensional approximation, are shown in Figs. 5.6 and 5.7, respectively.

5.4   Global-Local Orthogonal Approximation in 1-, 2- and N-Dimensional Spaces

In the previous section we discussed a novel weighting function interpolation and approximation technique to blend arbitrary smooth overlapping local functional approximations. The weight functions are designed to guarantee the global continuity conditions while retaining near complete freedom on the selection of the generating local approximations. Of course the key to the success of the proposed method depends upon the approximation capa-

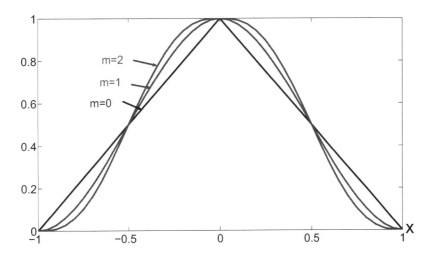

FIGURE 5.6
1-D Weighting Functions for Various Degrees of Piecewise Continuity.

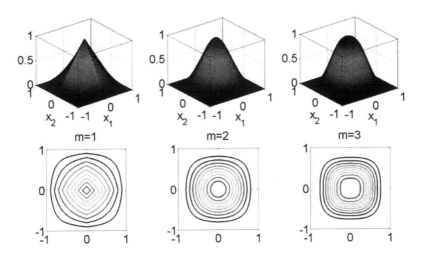

FIGURE 5.7
2-D Weighting Functions for Various Degrees of Piecewise Continuity.

TABLE 5.1
Weight Functions for Higher Order Continuity

Order of piecewise continuity	Weight Function: $$w_{0,0,\cdots.0}(x_1, x_2, \cdots, x_N) = \prod_{i=1}^{N} w(x_i)$$ $w(x)$, for all $x \in \{-1 \le x \le 1\}$, $y \overset{\Delta}{=} |x|$
0	$w(x) = 1 - y$
1	$w(x) = 1 - y^2(3 - 2y)$
2	$w(x) = 1 - y^3(10 - 15y + 6y^2)$
3	$w(x) = 1 - y^4(35 - 84y + 70y^2 - 20y^3)$
$\vdots$	$\vdots$
m	$w(x) = 1 - y^{m+1}\left\{ \frac{(2m+1)!(-1)^m}{(m!)^2} \sum_{k=0}^{m} \frac{(-1)^k}{2m-k+1} \binom{m}{k} y^{m-k} \right\}$

bility of the local functions. As can be seen from the first four chapters of this text, there are infinitely many ways to specify good preliminary local approximations averaged in Eq. (5.25). Ideally, one would like to choose these preliminary local approximations based upon some a priori information that we may have about the input-output data. Alternatively, these local models can also be constructed intelligently by studying local behavior of an input-output map by the use of methods like principal component analysis or Fourier decomposition [58–60]. Such special functions may be introduced either by themselves or to supplement a previously existing approximation. This aspect, called "basis enrichment" in the FEM literature, is one of the greatest advantages of this new generalized framework because it provides the invaluable option of very easily using local functions of different form and number in the individual sub-domains, based solely on local approximability. If no a priori information is available about the input-output data model, then one can choose these local models as a linear combination of basis functions taken from a large dictionary [46], and sparsity criteria can be used to find the least number of coefficients with careful attention to the approximation errors. More details about this procedure can be found in Refs. [45, 46]. In particular, if one chooses these basis functions to be a set of orthogonal polynomials, then Fourier coefficients of each preliminary approximation can be efficiently computed from ratios of inner products, avoiding any matrix inversion. Furthermore, Fourier coefficients corresponding to each basis function are independent of each other, and so inclusion of a new basis function in the basis vector does not require us to re-solve for previously computed Fourier coefficients. In this section, we illustrate the procedure of computing the preliminary approximations by using some novel orthogonal basis functions.

5.4.1   1-Dimensional Case

Consider the approximation of a one variable function $F(X)$. Suppose we are using the weighting function method as illustrated in Fig. 5.3. The preliminary approximations, while arbitrary, could be chosen to minimize the least squares criterion

$$J = \frac{1}{2} \int_{-1}^{1} w(x)[F(X) - F_I(X)]^2 dx \tag{5.29}$$

Note that the $w(x)$ in Eq. (5.29) defines the region of validity for the local approximation $F_I(X)$, or in other words, it provides the compact support to local approximation $F_I(X)$. While Eq. (5.29) holds for any positive $w(x)$, we consider in particular the weight functions of Table 5.1 associated with the weighted average of Eq. (5.1). Furthermore, we consider the case that $F_I(X)$ is a linear combination of any prescribed set of linearly independent basis functions $\{\phi_0(x), \phi_1(x), \cdots, \phi_n(x)\}$ as

$$F_I(X) = \sum_{i=0}^{n} a_i \phi_i(x); \; X = \,^I X + hx \tag{5.30}$$

The least squares criterion, making use of Eq. (5.29), can be written as

$$J = J_0 - c^T a + \frac{1}{2} a^T M a \tag{5.31}$$

where

$$J_0 = \frac{1}{2} \int_{-1}^{1} w(x) F^2(x) dx \equiv \frac{1}{2} < F(x), F(x) > \tag{5.32}$$

$$c^T = \left\{ \int_{-1}^{1} w(x)F(x)\phi_0(x)dx \int_{-1}^{1} w(x)F(x)\phi_1(x)dx \cdots \int_{-1}^{1} w(x)F(x)\phi_n(x)dx \right\}$$

$$\equiv \left\{ < F(x), \phi_0(x) > < F(x), \phi_1(x) > \cdots < F(x), \phi_n(x) > \right\} \tag{5.33}$$

$$M = \begin{bmatrix} \mu_{00} & \mu_{01} & \cdots & \mu_{0n} \\ \mu_{01} & \mu_{11} & \cdots & \mu_{1n} \\ \vdots & \vdots & \ddots & \vdots \\ \mu_{0n} & \mu_{1n} & \cdots & \mu_{nn} \end{bmatrix} = M^T \tag{5.34}$$

$$a \equiv \left\{ a_0 \; a_1 \cdots a_n \right\}^T; \; \mu_{ij} = < \phi_i, \phi_j > \equiv \int_{-1}^{1} w(x)\phi_i(x)\phi_j(x)dx \tag{5.35}$$

Observe that minimization of Eq.(5.31) gives the optimum (minimum integral least squares fit error) coefficients as

$$a = M^{-1} c \tag{5.36}$$

While Eq. (5.36) holds for an arbitrary set of linearly independent basis functions, for the special case that the basis functions satisfy the orthogonality condition

$$< \phi_i(x), \phi_j(x) > \equiv \int_{-1}^{1} w(x)\phi_i(x)\phi_j(x)dx = k_i\delta_{ij}, \; k_i \overset{\Delta}{=} \mu_{ii} = \int_{-1}^{1} w(x)\phi_i^2(x)dx,$$

(5.37)

the least squares solution of Eq. (5.36), as a consequence of the diagonal M matrix, simplifies to the simple uncoupled result to compute the Fourier coefficients:

$$a_i = \frac{< F(x), \phi_i(x) >}{k_i}, \; i = 1, 2, \cdots, n$$

(5.38)

Thus, if we can construct basis functions orthogonal with respect to the particular weight functions of Table 5.1, we enjoy the usual advantages that flow from approximation of orthogonal functions but now in a global/local approximation setting.

We consider the special case of $m = 1$; the construction of the corresponding orthogonal basis functions requires the Gramm-Schmidt process or the hypergeometric differential equation process. Using the methods from Chapter 2, it can be verified that the basis functions given in Table 5.2 satisfy the orthogonality conditions of Eq. (5.37). Note that c_n in Table 5.2 is determined so that $\phi_n(x)$ satisfies the normalization, $|\phi_n(\pm 1)| = 1$. The first four orthogonal functions are plotted in Fig. 5.8.

Qualitatively, we have realized a remarkable multi-resolution approach which combines the power of orthogonal approximation with the FEM. Furthermore, we achieved this capability in a framework that readily generalizes to n dimensions as we show below.

5.4.2 2-Dimensional Case

Consider the approximation of a two-variable function $F(X_1, X_2)$. The typical preliminary local approximations $F_{IJ}(X_1, X_2)$, while arbitrary, in particular could be chosen to minimize the least squares criterion

$$J = \frac{1}{2} \int_{-1}^{1} \int_{-1}^{1} w(x_1, x_2)[F(X_1, X_2) - F_{IJ}(X_1, X_2)]^2 dx_1 dx_2$$

(5.39)

Furthermore, we consider the case that $F_{IJ}(X_1, X_2)$ is chosen as a linear combination of a prescribed set of linearly independent basis functions $\{\phi_{ij}(x)\}$; $i = 1, 2, ...n; j = 1, 2, ...n$ as

$$F_{IJ}(X_1, X_2) = \sum_{i=0}^{n} \sum_{j=0}^{n} a_{ij}\phi_{ij}(x_1, x_2); \; X_1 = {}^{I}X_1 + hx_1, \; X_2 = {}^{J}X_2 + hx_2$$

(5.40)

TABLE 5.2

1-Dimensional Basis Functions Orthogonal with Respect to the Weight
Function $(x) = 1 - x^2(3 - 2|x|)$

Degree	Basis Functions, $\phi_j(x)$
0	1
1	x
2	$(-2 + 15x^2)/13$
3	$(-9x + 28\,x^3)/19$
$\vdots$	$\vdots$
n	$\phi_n(x) = \frac{1}{c_n}\left[x^n - \sum\limits_{j=0}^{n-1} \frac{<x^n, \phi_j(x)>}{<\phi_j(x), \phi_j(x)>}\phi_j(x)\right]$

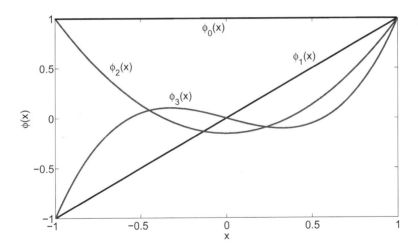

FIGURE 5.8

1-Dimensional Orthogonal Basis Functions.

In particular, consider the multiplicative structure for the weight function

$$w(x_1, x_2) = [1 - x_1^2(3 - 2|x_1|)][1 - x_2^2(3 - 2|x_2|)] \tag{5.41}$$

and the corresponding 2-dimensional basis functions

$$\phi_{ij}(x_1, x_2) = \phi_i(x_1)\phi_j(x_2) \tag{5.42}$$

where we choose the 1-dimensional basis functions $\phi_i(x)$ from Table 5.2 that are orthogonal with respect to $w(x) = 1 - x^2(3 - 2|x|)$. Introducing the inner product notation

$$< \alpha(x_1, x_2), \beta(x_1, x_2) > \overset{\Delta}{=} \int\limits_{-1}^{1} \int\limits_{-1}^{1} w(\xi_1, \xi_2)\alpha(\xi_1, \xi_2)\beta(\xi_1, \xi_2)d\xi_1 d\xi_2 \tag{5.43}$$

As a consequence of the orthogonality $\phi_i(x)$ from Table 5.2, the choice of Eqs. (5.41) and (5.42), and the definition of Eq. (5.43), it is evident that the functions of Eqs. (5.41) are orthogonal because

$$< \phi_{ij}(x_1, x_2), \phi_{lm}(x_1, x_2) > \overset{\Delta}{=} \int\limits_{-1}^{1} \int\limits_{-1}^{1} w(\xi_1)w(\xi_2)\phi_{ij}(\xi_1, \xi_2), \phi_{lm}(\xi_1, \xi_2)d\xi_1 d\xi_2$$

$$= \underbrace{\int\limits_{-1}^{1} w(\xi_1)\phi_i(\xi_1)\phi_l(\xi_1)d\xi_1}_{k_i \delta_{il}} \underbrace{\int\limits_{-1}^{1} w(\xi_2)\phi_j(\xi_2)\phi_m(\xi_2)d\xi_2}_{k_j \delta_{jm}}$$

$$= k_i \delta_{il} k_j \delta_{jm}$$

$$\tag{5.44}$$

Thus, remarkably, the 1-dimensional orthogonality properties directly generate the corresponding 2-dimensional orthogonality condition. As a consequence of orthogonality, it follows that the Fourier coefficients for 2-dimensional approximations are:

$$a_{ij} = \frac{< \phi_{ij}(x_1, x_2), F(X_1, X_2) >}{< \phi_{ij}(x_1, x_2), \phi_{ij}(x_1, x_2) >} = \frac{< \phi_{ij}(x_1, x_2), F(X_1, X_2) >}{k_i k_j} \tag{5.45}$$

The first five sets (degrees zero through four) of the 2-dimensional orthogonal polynomials of Eq. (5.42) are shown in Fig. 5.9.

5.4.3   *N*-Dimensional Case

Consider the approximation of a function $F(X_1, X_2, \cdots, X_N)$ of N variables. The preliminary local approximations $F_{I_1 \cdots I_N} F(X_1, X_2, \cdots, X_N)$, while arbitrary, in particular could be chosen to minimize the least squares criterion:

$$J = \frac{1}{2} \int\limits_{-1}^{1} \cdots \int\limits_{-1}^{1} w(x_1, \cdots, x_N) [F(X_1, \cdots, X_N) -$$

$$F_{I_1 \cdots I_N}(X_1, \cdots, X_N)]^2 dx_1 \cdots dx_N \tag{5.46}$$

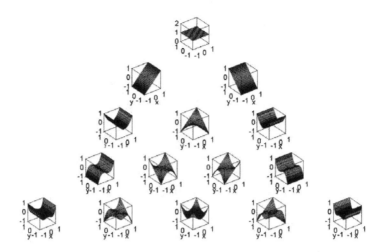

FIGURE 5.9
2-Dimensional Orthogonal Basis Functions.

Furthermore, we consider the case that $F_{I_1\cdots I_N}F(X_1, X_2, \cdots, X_N)$ is chosen as a linear combination of a prescribed set of linearly independent basis functions $\{\phi_{i_1\cdots i_N}(x_1, \cdots, x_N)\}$ as

$$F_{I_1\cdots I_N}(X_1, \cdots, X_N) = \sum_{i_1=0}^{n} \cdots \sum_{i_N=0}^{n} a_{i_1\cdots i_N} \phi_{i_1\cdots i_N}(x_1, \cdots, x_N) \qquad (5.47)$$

where the transformation from the global coordinates $(X_1, \cdots, X_N)$ to the local non-dimensional coordinates $(x_1, \cdots, x_N)$ is

$$X_1 = {}^{I_1}X_1 + hx_1, \cdots, X_N = {}^{I_N}X_N + hx_N \qquad (5.48)$$

In particular, consider the continued product structure for the weight function

$$w(x_1, \cdots, x_N) = [1 - x_1^2(3 - 2|x_1|)] \cdots [1 - x_N^2(3 - 2|x_N|)] = \prod_{i=0}^{N}[1 - x_i^2(3 - 2|x_i|)] \qquad (5.49)$$

and similarly for the basis functions

$$\phi_{i_1\cdots i_N}(x_1, \cdots, x_N) = \prod_{i=0}^{N} \phi_i(x_i) \qquad (5.50)$$

where we choose the 1-dimensional basis functions $\phi_i(x)$ from Table 5.2 that are orthogonal with respect to $w(x) = 1 - x^2(3 - 2|x|)$.

Introducing the inner product notation:

$$< \alpha(x_1, \cdots, x_N), \beta(x_1, \cdots, x_N) > \overset{\Delta}{=}$$

$$\int\limits_{-1}^{1} \cdots \int\limits_{-1}^{1} w(\xi_1, \cdots, \xi_N) \alpha(\xi_1, \cdots, \xi_N) \beta(\xi_1, \cdots, \xi_N) d\xi_1 \cdots d\xi_N \quad (5.51)$$

As a consequence of the orthogonality $\phi_i(x)$ from Table 5.2, and the choice of Eqs. (5.46), (5.49) and (5.50), it is evident that the N-dimensional functions of Eqs. (5.47) are orthogonal because

$$< \phi_{i_1 \cdots i_N}(x_1, \cdots x_N), \phi_{j_1 \cdots j_N}(x_1, \cdots x_N) >$$

$$= \int\limits_{-1}^{1} \cdots \int\limits_{-1}^{1} \left(w(\xi_1) \cdots w(\xi_N) \phi_{i_1 \cdots i_N}(\xi_1, \cdots, \xi_N) \phi_{j_1 \cdots j_N}(\xi_1, \cdots, \xi_N) \right) d\xi_1 \cdots d\xi_N$$

$$= [k_{i_1} k_{i_2} \cdots k_{i_N}][\delta_{i_1 j_1} \delta_{i_2 j_2} \cdots \delta_{i_N j_N}] \quad (5.52)$$

As a consequence of orthogonality, it follows that the least squares amplitudes are the Fourier coefficients,

$$a_{i_1 \cdots i_N} = \frac{< \phi_{i_1 \cdots i_N}(x_1, \cdots, x_N), F(X_1, \cdots, X_N) >}{\prod\limits_{j=1}^{N} k_j} \quad (5.53)$$

Remarkably, the weight functions, basis functions and orthogonality conditions for N-dimensional approximation are generated directly from the 1-dimensional results. Furthermore, we arrive at a computationally efficient piecewise continuous approximation in N dimensions with the added benefits that the local approximations are linear combinations of basis functions orthogonal to the same weight function used in averaging the overlapping approximations. Clearly, the first-order piecewise continuity implicit in the above developments is "promoted" from first to m^{th} order continuity by simply choosing the appropriate weight function from Table 5.1.

5.5   Algorithm Implementation

In this section, the step-by-step implementation of the Global-Local Orthogonal Mapping (GLO-MAP) algorithm is discussed. Attention here is upon the hyper surface approximation when unevenly spaced discrete measured data are available, whereas the above developments are for the case of continuous measurements. The main steps of the GLO-MAP algorithm are as follows:

1. Choose a set of sequential neighboring points, ^IX, arbitrary in number and location. These points serve as the *centroids of validity* for the local functional approximation, F_I. The density and location of these points depends upon numerous factors, like location and density of available measurement data and desired degree of approximation.

2. Choose a set of basis functions based on computational efficiency or a priori knowledge of the nature of the given input-output data to approximate $F(X)$ in the local neighborhood of a centroid of validity, ^IX. The local neighborhood of a centroid is defined in a such a way that the number of measurement points in the local neighborhood are at least equal to the number of basis functions used to approximate local behavior in that particular local domain. Generally, the sizing of the local neighborhood is dictated by the support or domain of the weight functions discussed in Section 5.3. One attractive choice for the basis functions is the orthogonal polynomial basis functions as discussed in the previous section.

3. Determine the Fourier coefficients corresponding to each local approximation. For the approximation of a given continuous functional form, or from dense discretely measurable functions, the Fourier coefficients can be computed by numerically evaluating the integral expression in Eq. (5.53) using standard numerical integration algorithms [71]. In the case of the GLO-MAP algorithm, the numerical integration procedure can be summarized as below:

   (a) Determine Gaussian quadrature points, $\mathbf{X}_G$, in unit hypercube.

   (b) For each quadrature point, determine the measurement points, $\mathbf{X}_i$, for which the weight function has a non-zero value.

   (c) Determine function value at the quadrature point, $F(\mathbf{X}_G)$, which can be taken as a weighted average of function value at $\mathbf{X}_i$.

$$F(\mathbf{X}_G) = \frac{1}{\sum_i w(\mathbf{X}_i)} \sum_i F(\mathbf{X}_i)w(\mathbf{X}_i) \qquad (5.54)$$

   (d) Evaluate the numerical integral in Eq. (5.53).

It should be noted that Gaussian quadrature points are the same for each local approximation and can be pre-computed. However, the numerical evaluation of integral expression of several variables, over regions with dimension greater than one, is not easy. As a rule of thumb, the number of function evaluations needed to sample an N-dimensional space increases as the Nth power of the number needed to do a 1-dimensional integral, resulting in increased computational cost associated with numerical integration proportion to the Nth power. To avoid these difficulties, one can construct the orthogonal basis functions which satisfy exactly the discrete orthogonality condition, using the hypergeometric difference equation and the procedure given in Ref. [12].

Discrete Orthogonality Condition: $\sum_i \phi_n(X_i)w(X_i)\phi_m(X_i) = k_{mn}^2 \delta_{mn}$

$$(5.55)$$

However, the major drawback of constructing discrete orthogonal polynomials is that their functional form changes for each local approximation depending upon the number of measurements available in each local neighborhood. More generally, and especially for a general not-necessarily dense set of discrete measurements, conventional SVD or linear least squares algorithms can be employed to construct the local approximations. It should be noted that the freedom to select local approximations affords a new level of flexibility in numerical methods for the class of problems under consideration.

4. The final step of the GLO-MAP algorithm is the use of weighting functions to merge the local approximations into a single m^{th} order continuous functional model. This is accomplished by using the weight functions listed in Table 5.1 and Eqs. (5.25)−(5.27). Note this step is the most important feature of the GLO-MAP algorithm as it reduces the systematic error introduced due to the neglected interaction between different local models by blending overlapping local approximations into a global one.

Note, as usual, that the size, h, of the local neighborhood is an important factor which affects the overall accuracy and computational cost of the GLO-MAP algorithm. To find the optimal value of this parameter, analogous to mesh refinement in the FEM method, one can construct a multi-resolution algorithm which iteratively refines the local neighborhoods until introduction of more local neighborhoods does not bring any improvement in the learning of input-output mapping. The major steps involved in this algorithm are depicted in Fig. 5.10.

5.5.1   Sequential Version of the GLO-MAP Algorithm

There are many engineering application problems that need to be solved in an iterative manner in real time by successively approximating the input-output data. Many recursive approximation algorithms are presented in the literature, but the Kalman filter [2] is one of the most widely used and powerful tools for recursive estimation problems. We note that the Kalman filter algorithm is very attractive for the problem at hand as it can also be used to update off-line a priori learned GLO-MAP network parameters in real time whenever new measurements are available. However, the main challenge associated with the use of the Kalman filter in the GLO-MAP algorithm is the dynamic state vector, i.e., the components of the state vector of the Kalman filter changes

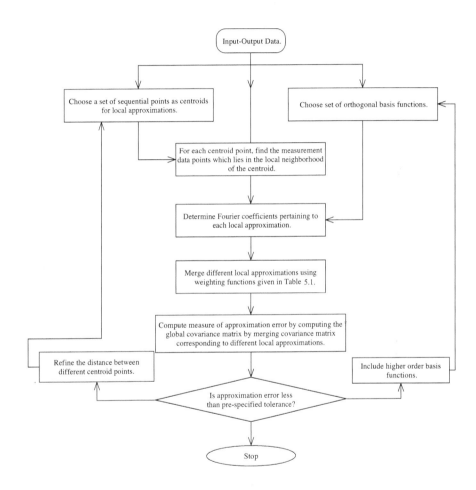

FIGURE 5.10
Flowchart for the GLO-MAP-Based Multi-Resolution Algorithm.

with every measurement data depending upon the location of measurement data relative to centroids of validity of local approximations. Actually, the total number of unknowns for the GLO-MAP network is Mn coefficients of different local approximations. Here, n is the number of basis functions associated with each local approximation, and M is the total number of these local approximations. However, when a new measurement is available, we just need to update $2^N n$ unknowns depending upon the location of measurement data, since only the neighboring 2^N local approximations (associated with the 2^N vertices of the hypercube containing the new measurement) will have non-zero contribution in the final global map obtained by merging different local approximations using Eq. (5.25). The main steps involved in the implementation of the sequential version of the GLO-MAP algorithm are as follows:

1. Depending upon prior knowledge of the input space, choose a set of sequential points, $^I X$, which serve as the centroids of validity for the local functional approximations, F_I. The density and location of these points depend upon numerous factors like location and density of available measurement data and desired degree of approximation.

2. Choose a set of basis functions (preferably orthogonal functions) to approximate $F(X)$ in the local neighborhood of a centroid of validity, $^I X$. Initialize the coefficients of each basis function to be zero, and the associated covariance matrix $(\mathbf{P_G})$ to be identity times a large number.

3. Given a new measurement data point, $(\mathbf{X}, F(\mathbf{X}))$, find the neighboring 2^N centroids such that the weight function associated with these 2^N centroids have a non-zero value at the measurement point.

   (a) Include the coefficients of the basis functions associated with these centroids in the local algebraic Kalman filter state vector, denoted by $\mathbf{x}$. Let $\mathbf{x}_G$ denote the global super-set of coefficients and $\mathbf{M}$ be a *selection matrix* consisting of zeros and ones that satisfies $\mathbf{x} = \mathbf{M}\mathbf{x}_G$.

   (b) Extract rows and columns of $\mathbf{P}_G$ corresponding to the coefficients associated with these 2^N centroids and denote them by $\mathbf{P}_k^- = \mathbf{M}\mathbf{P}_G\mathbf{M}^T$.

4. Use the following equations to compute the local Kalman gain and update state vector, $\mathbf{x}$, and the associated covariance matrix, $\mathbf{P}$,

$$\mathbf{x}_k^+ = \mathbf{x}_k^- + \mathbf{K}(F(\mathbf{X}) - \mathbf{\Phi}_k(.)\mathbf{x}_k^-) \tag{5.56}$$

$$\mathbf{P}_k^+ = (\mathbf{I} - \mathbf{K}\mathbf{\Phi}_k(.))\mathbf{P}_k^- \tag{5.57}$$

$$\mathbf{K} = \mathbf{P}_k^- \mathbf{\Phi}_k(.)^T (\mathbf{\Phi}_k(\mathbf{X})\mathbf{P}_k^- \mathbf{\Phi}_k(.)^T + \mathbf{R}_k)^{-1} \tag{5.58}$$

where superscript $^-$ and $^+$ denote the value of variables before and after updating various unknowns using given measurement data, respectively.

Subscript k denotes the centroid number associated with the kth approximation and varies from 1 to 2^N. Further, matrix $\mathbf{\Phi}$ and $\mathbf{R}$ are given by the following equations:

$$\mathbf{\Phi}_k = \begin{bmatrix} \phi_1(\mathbf{X}_k) \cdots \phi_N(\mathbf{X}_k) \end{bmatrix} \tag{5.59}$$

$$\mathbf{R}_k = \sigma w(\mathbf{X}_k) \tag{5.60}$$

where $\mathbf{X}_k$ denotes the local coordinates of a measurement point assuming origin at the kth neighboring centroid and σ denotes the variance of the measurement data.

5. Update rows and columns of the global covariance matrix, $\mathbf{P}_G$, and coefficients associated with each local approximation.

6. Once the value of $\mathbf{P}_G$ is less than a pre-specified tolerance, use appropriate weighting functions to merge various local approximations into a single m^{th} order continuous functional model. This is accomplished by using the weight functions listed in Table 5.1 and Eqs. (5.25)–(5.27), centered at the 2^N vertices of the local volume containing the point $\mathbf{x}$.

5.6 Properties of GLO-MAP Approximation

5.6.1 Approximation Error

One of the key advantages of least squares approximations lies in the fact that residual error e is orthogonal to the range space spanned by basis functions $\{\phi_i\}$. Now, we state the following lemma according to which the residual error, after carrying out the GLO-MAP averaging process, is orthogonal to the range space spanned by orthogonal basis functions, $\phi_i(x)$.

Lemma 4. *Let $\{\phi_i\}$ be a set of orthogonal polynomials used in the GLO-MAP algorithm with respect to the weight function, w, over interval $[-1, 1]$ where subscript i denotes the degree of the polynomial. Let $\hat{f}$ denote the GLO-MAP approximation of a continuous function f over the interval $[a, b]$*

$$\hat{f} = \sum_{i=1}^{m} w_i \hat{f}_i \tag{5.61}$$

where m is the total number of local approximations, $\hat{f}_i$ is the local least squares approximation in the ith interval and w_i is the specially designed GLO-MAP weight function associated with the ith interval. Now, if we define residual error $e = f - \hat{f}$ then e is orthogonal to the range space spanned by the basis functions, ϕ_i, under the norm induced by $\sum_{i=1}^{m} w_i(x) = 1$.

Proof. Note as each local approximation, $\hat{f}_i$, denotes the least squares approximation over the ith interval, the following holds:

$$\int\limits_{x_{il}}^{x_{iu}} (f(x) - \hat{f}_i(x, x_i))w_i(x, x_i)\phi_j(x)dx = \int\limits_a^b (f(x) - \hat{f}_i(x, x_i))w_i(x, x_i)\phi_j(x)dx$$

$$= 0, \ j = 1, 2, \cdots, n \qquad (5.62)$$

where x_{il} and x_{iu} denote the lower and upper limits of the ith local interval. Now, let us consider the residual error over the whole interval $[a, b]$

$$\int\limits_a^b (f(x) - \hat{f}(x))\phi_j(x)dx = \int\limits_a^b f(x)\phi_j(x)dx - \sum_{i=1}^m \int\limits_a^b w_i(x, x_i)\hat{f}_i(x, x_i)\phi_j(x)dx,$$

$$j = 1, 2, \cdots, n \qquad (5.63)$$

Further, from Eqs. (5.62) and (5.63), we have

$$\int\limits_a^b (f(x) - \hat{f})\phi_j dx = \int\limits_a^b f(x)\phi_j(x)dx - \sum_{i=1}^m \int\limits_a^b w_i(x, x_i)f(x)\phi_j(x)dx,$$

$$j = 1, 2, \cdots, n \quad (5.64)$$

Now, as the GLO-MAP weight functions satisfy the *partition of unity* paradigm, Eq. (5.64) reduces to

$$\int\limits_a^b (f(x) - \hat{f})\phi_j dx = \int\limits_a^b f(x)\phi_j(x)dx - \int\limits_a^b f(x)\phi_j(x)dx = 0, \ j = 1, 2, \cdots, n$$

$$(5.65)$$

Hence, the GLO-MAP residual error, e, is orthogonal to range space spanned by the basis functions, ϕ_i. □

Although we have shown that the GLO-MAP residual error, e, is orthogonal to the range space spanned by the basis functions, ϕ_i, Theorem 3 does not hold for the final GLO-MAP approximation. This is due to the fact that the orthogonality condition of the basis functions and the orthogonality of the residual error, e, to the range space are not induced by same norm. Also, obviously, the final GLO-MAP approximation polynomials are of higher degree than the generating polynomials since the weight functions themselves are of degree $m + 2$, where m is the degree of piecewise continuity desired.

5.6.2   Bounds on Approximation Error

In this section, we state the following theorem to have a measure of net approximation error after merging different local approximations using the GLO-MAP algorithm.

Theorem 5. *Let f be a continuous function over N-dimensional space $\Omega \subset \mathcal{R}^N$. Let $\{w_i\}$ be a set of specially designed GLO-MAP weight functions with compact support Ω_i satisfying*

1. *$\{\Omega_i\}$ is an open cover for Ω*

2. *$\sum\limits_i w_i = 1$ on Ω*

3. *$\|w_i\|_\infty \leq 1$*

4. *$\|\nabla w_i\|_\infty \leq \frac{1}{h_i}$*

where h_i denotes the size of the ith sub-domain Ω_i. Assume that $\hat{f}_i$ denotes the approximation of f on sub-domain Ω_i such that

$$\|f - \hat{f}_i\|_{\mathcal{L}_2(\Omega_i)} \leq e_{1_i} \tag{5.66}$$

$$\|\nabla(f - \hat{f}_i)\|_{\mathcal{L}_2(\Omega_i)} \leq e_{2_i} \tag{5.67}$$

Then the GLO-MAP approximation $\hat{f} = \sum\limits_i w_i \hat{f}_i$ satisfies the following error bounds:

$$\|f - \hat{f}\|_{\mathcal{L}_2(\Omega)} \leq 2^{\frac{N}{2}} \left(\sum_i e_{1_i}^2 \right)^{\frac{1}{2}} \tag{5.68}$$

$$\|\nabla(f - \hat{f})\|_{\mathcal{L}_2(\Omega)} \leq 2^{\frac{N+1}{2}} \left(\sum_i \frac{e_{1_i}^2}{h_i^2} + \sum_i e_{2_i}^2 \right)^{\frac{1}{2}} \tag{5.69}$$

Proof. Since the weight functions, w_i, form a partition of unity over Ω, we have

$$f = 1.f = \sum_i w_i f \tag{5.70}$$

Substituting for f from Eq. (5.70) in Eq. (5.68), we get

$$\|f - \hat{f}\|_{\mathcal{L}_2(\Omega)}^2 = \| \sum_i w_i(f - \hat{f}_i)\|_{\mathcal{L}_2(\Omega)}^2 \tag{5.71}$$

Since, at any point $\mathbf{x} \in \Omega$ only 2^N local approximations overlap, the summation terms in Eq. (5.71) also contain at most 2^N terms for any $x \in \Omega$. Thus, we have

$$\|f - \hat{f}\|_{\mathcal{L}_2(\Omega)}^2 \leq 2^N \sum_i \|w_i(f - \hat{f}_i)\|_{\mathcal{L}_2(\Omega)}^2 \tag{5.72}$$

Further, making use of the fact that the support of weight function w_i is Ω_i, we have

$$\|f - \hat{f}\|^2_{\mathcal{L}_2(\Omega)} \leq 2^N \sum_i \|w_i(f - \hat{f}_i)\|^2_{\mathcal{L}_2(\Omega)}$$

$$\leq 2^N \sum_i \|w_i(f - \hat{f}_i)\|^2_{\mathcal{L}_2(\Omega)_i}$$

$$\leq 2^N \sum_i 1.e^2_{1_i} = 2^N \sum_i e^2_{1_i} \qquad (5.73)$$

This proves the estimates of Eq. (5.68). To show the estimates of Eq. (5.69), let us consider $\|\nabla(f - \hat{f})\|^2_{\mathcal{L}_2(\Omega)}$

$$\|\nabla(f - \hat{f})\|^2_{\mathcal{L}_2(\Omega)} = \|\nabla(f - \hat{f}_i)\|^2_{\mathcal{L}_2(\Omega)}$$

$$\leq \|\nabla \sum_i w_i(f - \hat{f}_i)\|^2_{\mathcal{L}_2(\Omega)}$$

$$\leq 2\|\sum_i \nabla w_i(f - \hat{f}_i)\|^2_{\mathcal{L}_2(\Omega)} + 2\|\sum_i w_i \nabla(f - \hat{f}_i)\|^2_{\mathcal{L}_2(\Omega)}$$

$$\leq 2^{N+1} \sum_i \left(\|\nabla w_i(f - \hat{f}_i)\|^2_{\mathcal{L}_2(\Omega_i)} + \|w_i \nabla(f - \hat{f}_i)\|^2_{\mathcal{L}_2(\Omega_i)} \right)$$

$$\leq 2^{N+1} \sum_i \left(\frac{1}{h^2_i} e^2_{1_i} + e^2_{2_i} \right)$$

which proves the theorem. □

Although Theorem 5 quantifies the approximation error of the GLO-MAP algorithm, it does not provide any information about the effect of the measurement error on the net approximation error. To quantify the effect of measurement error on the net approximation error, an alternative, probabilistic approach is presented in the next section.

5.6.3   Probabilistic Analysis of the GLO-MAP Algorithm

In estimating unknown parameters from a statistical model, one is interested in how the estimates deviate from the true value of the parameter. The deviations generally come from two sources.

1. Random Error: The source of this error is the random noise present in measurement data.

2. Bias or Systematic Error: Bias is the difference between the average value of the estimates from the true value.

The difference between the estimation algorithm bias error and the random error is that the estimate bias can typically be reduced by increasing the measurement data size, while the random error cannot be reduced arbitrarily. In this section, we quantify these errors for the GLO-MAP algorithm using a statistical approach and discuss some other statistical properties of the GLO-MAP algorithm.

According to the GLO-MAP procedure listed in the previous section, different local approximations are merged together using specially designed weight functions to obtain a desired order piecewise continuous global estimates

$$\hat{y}(\mathbf{x}) = \sum_{l=1}^{M} w(\mathbf{x}, \mathbf{x}_l)\hat{y}_l(\mathbf{x}_l) \tag{5.74}$$

where M is the total number of local approximations, and $\hat{y}_l = \boldsymbol{\phi}^T \mathbf{a}_l$ denotes the lth local approximation obtained by the least squares process. So, Eq. (5.74) can be rewritten as:

$$\hat{y}(\mathbf{x}) = \sum_{l=1}^{M} \underbrace{w(\mathbf{x}, \mathbf{x}_l)\mathbf{a}_l^T}_{\bar{\mathbf{a}}_l} \boldsymbol{\phi}(\mathbf{x}, \mathbf{x}_l) = \sum_{l=1}^{M} \bar{\mathbf{a}}_l^T \boldsymbol{\phi}(.) \tag{5.75}$$

$$= \boldsymbol{\Phi}^T \bar{\mathbf{a}} \tag{5.76}$$

where

$$\boldsymbol{\Phi} = \{\boldsymbol{\phi}(\mathbf{x}, \mathbf{x}_1), \cdots, \boldsymbol{\phi}(\mathbf{x}, \mathbf{x}_M)\}^T \tag{5.77}$$

$$\bar{\mathbf{a}} = \{\bar{\mathbf{a}}_1, \bar{\mathbf{a}}_2, \cdots, \bar{\mathbf{a}}_M\}^T \tag{5.78}$$

Now, using the linear error propagation theory, we can write

$$\mathbf{P}_{yy_G} = \boldsymbol{\Phi}^T \mathbf{P}_{\bar{a}\bar{a}} \boldsymbol{\Phi} \tag{5.79}$$

where $\mathbf{P}_{yy_G}$ denotes the global measurement estimate error covariance matrix, and $\mathbf{P}_{\bar{a}\bar{a}} = \mathcal{W}\mathbf{P}_{aa_G}\mathcal{W}$, with

$$\mathcal{W} = \begin{bmatrix} w(\mathbf{x}, \mathbf{x}_1)\mathbf{I}_{n \times n} & & \\ & \ddots & \\ & & w(\mathbf{x}, \mathbf{x}_M)\mathbf{I}_{n \times n} \end{bmatrix} \tag{5.80}$$

$$\mathbf{P}_{aa_G} = \begin{bmatrix} \mathbf{P}_{aa_1} & & \\ & \ddots & \\ & & \mathbf{P}_{aa_M} \end{bmatrix} \tag{5.81}$$

Note, $\mathcal{W}$ is a diagonal matrix with all entries less than or equal to one. As a consequence of this, $\mathbf{P}_{\bar{a}\bar{a}}$ can be regarded as a contraction mapping of $\mathbf{P}_{aa_G}$, i.e., we can write

$$\|\mathbf{P}_{\bar{a}\bar{a}}\| \leq \mathbf{P}_{aa_G} \tag{5.82}$$

Further, let us define the global measurement error covariance matrix, $\mathbf{P}_{yy}$, without the averaging process of the GLO-MAP algorithm:

$$\mathbf{P}_{yy} = \mathbf{\Phi}^T \mathbf{P}_{aa} \mathbf{\Phi} \tag{5.83}$$

Now, from Eqs. (5.79), (5.82) and (5.83), we can conclude that

$$\|\mathbf{P}_{yy_G}\| \leq \|\mathbf{P}_{yy}\| \tag{5.84}$$

Note Eq. (5.84) provides a quantitative justification for qualitative observation made earlier in this chapter: "If one least squares fit is good, the average of two should be better."

Finally, let us take the expected value of Eq. (5.74)

$$E[\hat{y}(\mathbf{x})] = E[\sum_{l=1}^{M} w(\mathbf{x}, \mathbf{x}_l)\hat{y}_l(\mathbf{x}_l)] \tag{5.85}$$

As local approximations $y_l(\mathbf{x}_l)$ are obtained by carrying out the least squares process, therefore, as a consequence of the unbiased property of the least squares estimator, we have

$$E[\hat{y}_l(\mathbf{x}_l)] = y(\mathbf{x}) \tag{5.86}$$

Now, substitution of Eq. (5.86) in Eq. (5.85) leads to the following equation for the expected value of the GLO-MAP approximation:

$$E[\hat{y}(\mathbf{x})] = \sum_{l=1}^{M} w(\mathbf{x}, \mathbf{x}_l) E[\hat{y}_l(\mathbf{x})] \tag{5.87}$$

$$= \sum_{l=1}^{M} w(\mathbf{x}, \mathbf{x}_l) y(\mathbf{x}) \tag{5.88}$$

Now, making use of the fact that weight functions $w(\mathbf{x}, \mathbf{x}_l)$ form a partition of unity, i.e., $\sum_{l=1}^{M} w(\mathbf{x}, \mathbf{x}_l) = 1$, we get

$$E[\hat{y}(\mathbf{x})] = \sum_{l=1}^{M} w(\mathbf{x}, \mathbf{x}_l) y(\mathbf{x}) = y(\mathbf{x}) \tag{5.89}$$

which proves that the GLO-MAP algorithm is an unbiased estimator. This is an important property to have, as in practice, unbiased estimators are rare. This truth, together with the covariance result of Eq. (5.84), provides a strong probabilistic justification of the GLO-MAP algorithm to augment the attractive localization and piecewise continuity features.

5.7   Illustrative Engineering Applications

The approximation algorithm presented in this chapter has been tested on a variety of engineering applications. In this section, we present four sets of results from these studies: (*i*) an analytical test case for function approximation; (*ii*) a dynamical system identification from wind tunnel testing of a synthetic jet actuation wing; (*iii*) a vibrating Space-Based Radar (SBR) antenna surface approximation; and (*iv*) an approximation of the "porkchop" surface for a family of Lambert's problem solutions.

5.7.1   Function Approximation

The test case for the analytical function approximation is constructed by using the following surface [38]:

$$f(X_1, X_2) = \frac{10}{(X_2 - X_1^2)^2 + (X_1 - 1)^2 + 1} + \frac{5}{(X_2 - 8)^2 + (X_1 - 5)^2 + 1}$$
$$+ \frac{5}{(X_2 - 8)^2 + (X_1 - 8)^2 + 1} \tag{5.90}$$

Figs. 5.12(a) and 5.12(b) show the true surface and contour plots for the function given by Eq. (5.90). According to our experience, this particular function has many important features, such as a sharp ridge line that is very difficult to learn with existing function approximation algorithms using a reasonable number of nodes.

To approximate the function given in Eq. (5.90), the whole input region is divided into a set of finite element cells, defined with Cartesian coordinates, X_1 and X_2. Therefore, the 10×10 modeling region can be divided into a different number of cells depending upon cell length. For example, the whole input region can be modeled by a total of 16 cells of dimension 2.5×2.5, or by a total of 576 cells of dimension 0.416×0.416. Further, to obtain preliminary approximations for a particular cell, as described in Section 5.4, two test cases are considered. In the first test case, the continuous functional expression given by Eq. (5.90) is used to obtain the coefficients of the preliminary local approximations, while in the second test case a discrete set of measurement data is used to compute preliminary approximations. The discrete measurement data set is generated by taking 100 random samples over the interval $[0\text{-}10, 0\text{-}10]$ for both X_1 and X_2, giving a total of 10^4 measurements.

The local approximation of analytical function, $\hat{f}(x_1, x_2)$, for a particular cell is modeled by orthogonal polynomials of the form:

$$\hat{f}(x_1, x_2) = \sum_i \sum_j a_{ij} \phi_i(x_1, x_2) \phi_j(x_1, x_2), \ i + j \le 2 \tag{5.91}$$

The orthogonal functions (listed in Table 5.2), ϕ_i and ϕ_j, are chosen in such a way that the degree of $\hat{f}(x_1, x_2)$ is always less than or equal to 2. Further, x_1 and x_2 denote the local cell coordinates defined as

$$x_1 = 2(X_1 - X_{1_m})/X_{1_{cell}} \quad x_2 = 2(X_2 - X_{2_m})/X_{2_{cell}} \tag{5.92}$$

where (X_{1_m}, X_{2_m}) and $X_{1_{cell}} \times X_{2_{cell}}$ represent the centroid and dimensions of a particular cell, respectively.

Since ϕ_i and ϕ_j are chosen to be orthogonal polynomial functions, unknown coefficients a_{ij} can be determined from Eq. (5.45). Due to the complex nature of the function in Eq. (5.90), the various integral expressions in Eq. (5.45) are computed by numerical integration as explained in Section 5.5. The total number of Gauss quadrature points required for numerical integration are decided by checking the orthogonality condition of Eq. (5.44), i.e., $< \phi_{ij}(x_1, x_2), \phi_{kl}(x_1, x_2) > \neq k_i \delta_{ik} k_j \delta_{jl}$. Fig. 5.11(a) shows the plot of orthogonalization error, defined as $|| < \phi_{ij}(x_1, x_2), \phi_{kl}(x_1, x_2) > - k_i \delta_{ik} k_j \delta_{jl} ||$, versus the number of Gauss quadrature points in a particular cell. As expected, the orthogonalization error decreases quickly as the number of quadrature points inside a particular cell increases. To generate this plot, we decided to use a total of 100 (10 in each direction) quadrature points.

Now, in the first test case, the analytical expression given by Eq. (5.90) is used to obtain the integrand values at different quadrature points, while in the second case the integrand values were obtained by a weighting average procedure, as discussed in Section 5.5. As discussed earlier, the approximation error depends upon the grid size; therefore, it was decided to study the Root Mean Square (RMS) approximation error as a function of cell size for a fixed order of polynomials. Fig. 5.11(b) shows the plot of root mean square approximation error versus cell size for both of the test cases. As expected, the root mean square error decreases for both test cases as cell size decreases. Due to the fact that as cell size decreases, the local behavior of the unknown function can be approximated more accurately, first order weighting function interpolation becomes more accurate. Further, it is also apparent from this figure that the RMS error for the first test case is less than for the second test case. This is because in the second test case the integrand values are obtained by interpolating the available measurement values in the local neighborhood of a particular quadrature point, while in first test case the analytical function expression is used to compute numerical integrals. It should also be noted that the the approximation accuracy for either test case is bounded by the orthogonalization error, which in this case is $O(10^{-5})$. These results provide a basis for optimism regarding the practical utility of this approach.

Finally, Figs. 5.13(a) and 5.13(b) show surface and contour plots of the approximated surface, whereas Figs. 5.13(c) and 5.13(d) show plots of the approximation error surface and the error contours, respectively. Not surprisingly, the largest errors occur along the knife edge of the sharp ridge; experimentation indicated that we can reduce the maximum error to any tolerance dictated by the feature sharpness and data spacing. These results

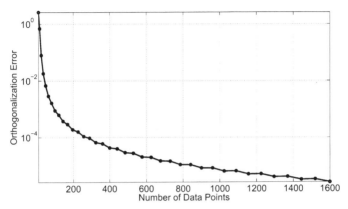

(a) Orthogonalization Error vs. Number of Data Points

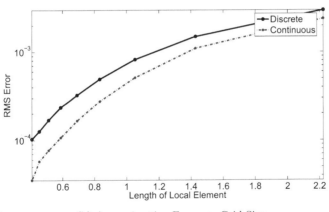

(b) Approximation Error vs. Grid Size

FIGURE 5.11
Error Analysis.

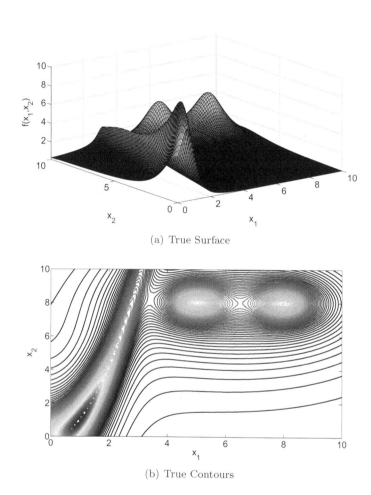

(a) True Surface

(b) True Contours

FIGURE 5.12

Test Surface and Contour Plots of Eq. (5.90).

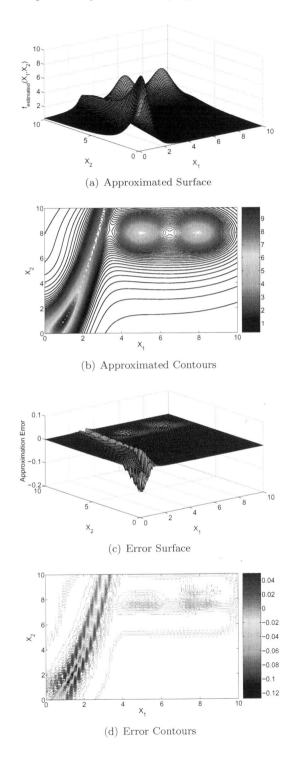

(a) Approximated Surface

(b) Approximated Contours

(c) Error Surface

(d) Error Contours

FIGURE 5.13
The GLO-MAP Approximation Results for Analytical Function of Eq. (5.90).

correspond to a total of 625 cells. From these figures, it is clear that we are able to learn the analytical function given in Eq. (5.90) very well with worst case relative approximation errors less than 2%. Of course, it is evident that using dense measurements along with either a smaller h or higher degree local approximations, we can make the errors as small as desired. As can be readily verified, this approximation is already superior to those of Chapters 2 and 4 using competing approaches.

We also mention that for any standard interpolation algorithm to represent the surface data shown in Fig. 5.12(a), the interpolation matrix consists of n^2 real numbers, or $\frac{n(n+1)}{2}$ real numbers. For example, if radial basis functions are used, we require $O(n^3)$ Floating Point Operations (FLOPs) to solve the associated system of equations. For a surface with 10^4 data points, one needs $O(10^{12})$ FLOPs, which is obviously impractical for many practical purposes. Further for any global representation, one needs to worry about the possible rank deficiency of the large interpolation matrix to solve the system of equations. However, the GLO-MAP algorithm proposed here greatly reduces the overall number of basis functions, requires no large matrix inverse, improves the surface approximation accuracy and provides a feasible path to achieve any desired precision by appropriate refinements. We now consider a diverse set of applications to obtain further insight.

5.7.2   Synthetic Jet Actuator Modeling

There is a significant, multifaceted effort in the aerospace industry to develop advanced technologies that would enable adaptive, intelligent, shape-controllable micro and macro structures for advanced aircraft and space systems. These designs involve precise control of the shape of the structures with micro- and macro-level manipulations (actuation). Synthetic jet actuators [72] (SJA) represent an alternative to reconfigurable wings that adaptively shape the flow and pressure fields around a fixed wing. This is one type of device being researched for active flow control that enables enhanced performance of conventional aerodynamic surfaces at high angles of attack, and these technologies may lead to full replacement of hinged control surfaces, thereby achieving hingeless control. Active flow control can be achieved by embedding sensors and actuators at micro scales on an aerodynamic structure. The desired force and moment profiles are achieved by impinging a jet of air to alter the flowfield using these actuators, thereby creating a desired pressure distribution over the structure. The distinguishing feature of the synthetic jet actuation modeling problem is that the relationship between input and output variables is poorly modeled and is nonlinear in nature. Further, unsteady flow effects make it impossible to capture the physics fully from static experiments. The issue at hand is to derive comprehensive mathematical models that capture the input-output behavior of SJAs so that one can derive automatic control laws that can command desired lift and moment profiles. While the conventional modeling approaches evolve to handle these problems, one can pursue non-parametric, multi-resolution, adaptive input-output modeling approaches to capture macro static and dynamic models directly from experiments. However, the large data sets need to be replaced by a consistent multi-dimensioned approximation, consistent with the accuracy of the measurement. In this section, we show the application of the GLO-MAP algorithm to learn the mapping between the synthetic jet actuation parameters (frequency, direction, etc., for each actuator) and the resulting aerodynamic lift, drag, and moment. These results show the effectiveness of the GLO-MAP algorithm presented in this chapter to learn the nonlinear input-output mapping for the synthetic jet actuation wing.

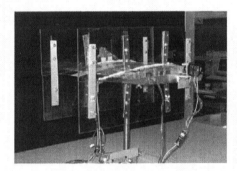

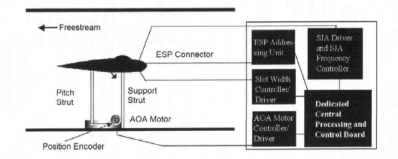

FIGURE 5.14
Hingeless Control-Dedicated Experimental Setup for Synthetic Jet Actuation Wing.

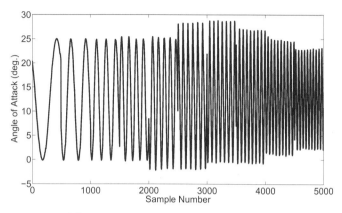

(a) Angle of Attack Variation without SJA.

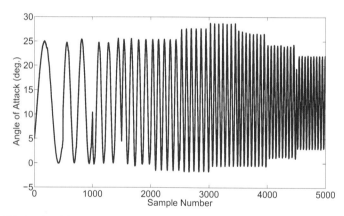

(b) Angle of Attack Variation with SJA Actuation Frequency of 60 Hz.

FIGURE 5.15

Angle of Attack Variation.

5.7.2.1   Experimental Setup

A hingeless-control-dedicated experimental setup has been developed. As part of the initial effort, a stand-alone control unit has been developed that controls all of the wing's and SJA's parameters and variables. The setup is installed in the $3' \times 4'$ wind tunnel of the Texas A&M Aerospace Engineering Department (Fig. 5.14*). The test wing profile for the dynamic pitch test of the synthetic jet actuator is an NACA 0015 airfoil. This shape was chosen due to the ease with which the wing could be manufactured and the available interior space for accommodating the synthetic jet actuator (SJA).

Experimental evidence [70, 72] suggests that a SJA, mounted such that its jet exits tangentially to the surface, has minimal effect on the global wing aerodynamics at low to moderate angles of attack. The primary effect of the jet is at high angles of attack when separation is present over the upper wing surface. In this case, the increased mixing associated with the action of a synthetic jet delays or suppresses flow separation. As such, the effect of the actuator is in the nonlinear post stall domain. To learn this nonlinear nature of SJA, experiments were conducted with the control-dedicated setup shown in Fig. 5.14. The wing Angle of Attack (AOA) is controlled by the following reference signal:

1. Oscillation type: sinusoidal oscillation magnitude: $12.5°$

2. Oscillation offset (mean AOA): $12.5°$

3. Oscillation frequency: from 0.2 Hz to 20 Hz

In other words, the AOA of the airfoil is forced to oscillate from $0°$ to $25°$ at a given frequency (see Fig. 5.15). The experimental data collected were the time histories of the pressure distribution on the wing surface (at 32 locations). The data was also integrated to generate the time histories of the lift coefficient and the pitching moment coefficient. Data were collected with the SJA on and with the SJA off (i.e., with and without active flow control). All the experimental data were taken for 5 sec at a 100 Hz sampling rate.

The experiments described above were performed at a free-stream velocity of 25 m/sec. From the surface pressure measurements, the lift and pitching moment coefficients were calculated via integration. The unknown SJA model is known to be dynamic in nature, so SJA wing lift force and pitching moment coefficients are modeled by a first-order system, i.e., they are assumed to be a function of current and previous time states (angle of attack).

$$C_L(t_k) = C_L(\alpha_{t_k}, C_L(t_{k-1})) \tag{5.93}$$
$$C_M(t_k) = C_M(\alpha_{t_k}, C_M(t_{k-1})) \tag{5.94}$$

---

*Please see the Appendix for the color version of this figure.

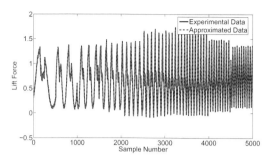

(a) 0 Hz Actuation Frequency

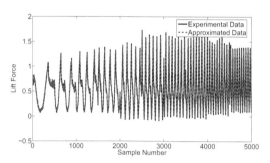

(b) 60 Hz Approximation Frequency

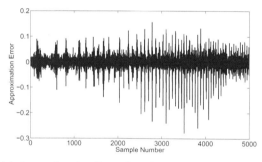

(c) Approximation Error for 0 Hz Actuation Frequency

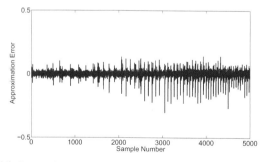

(d) Approximation Error for 60 Hz Actuation Frequency

FIGURE 5.16
Lift Force Approximation Results.

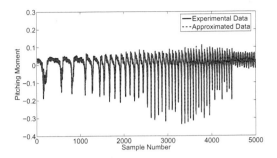

(a) 0 Hz Actuation Frequency

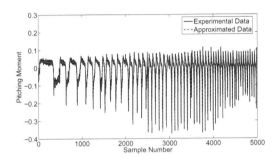

(b) 60 Hz Approximation Frequency

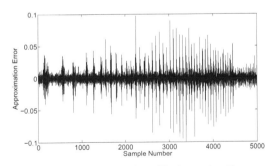

(c) Approximation Error for 0 Hz Actuation Frequency

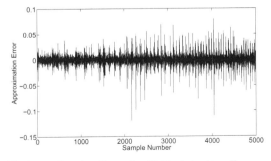

(d) Approximation Error for 60 Hz Actuation Frequency

FIGURE 5.17
Pitching Moment Approximation Results.

In this case, the moment and lift data are gridded based upon the time interval as described in the previous section. To approximate the dynamics in a particular time interval, the orthogonal functions, ϕ_{ij}, listed in Table 5.2 are used.

$$C_L(t_k) = \sum_i \sum_j a_{ij}\phi_i(\alpha(t_k))\phi_j(C_L(t_{k-1})), \ i+j \le 2$$

$$C_M(t_k) = \sum_i \sum_j a_{ij}\phi_i(\alpha(t_k))\phi_j(C_M(t_{k-1})), \ i+j \le 2 \qquad (5.95)$$

Figs. 5.16(a) and 5.16(b) show the measured and approximated lift coefficients for zero and 60 Hz jet actuation frequency, respectively, with a time interval size of 25. Figs. 5.16(c) and 5.16(d) show the corresponding approximation error plots. From these figures, it is clear that we are able to learn the nonlinear relationship between lift coefficient and angle of attack with and without SJA on.

Similarly, Figs. 5.17(a) and 5.17(b) show the measured and GLO-MAP approximated pitching moment coefficients for zero and 60 Hz jet actuation frequency, respectively. Figs. 5.17(c) and 5.17(d) show the corresponding approximation error plots. From these figures, it is clear that we are able to learn the nonlinear relationship between moment coefficient and angle of attack (with and without SJA being turned on) very well within experimental accuracy.

5.7.3 Space-Based Radar (SBR) Antenna Shape Approximation

Space-based radar systems envisioned for the future may be a constellation of large spacecraft that provide persistent real-time information of ground activities through the identification and tracking of moving targets, high-resolution synthetic aperture radar imaging and collection of high-resolution terrain information. The accuracy of the information obtained from SBR systems depends upon many parameters, like the geometric shape of the antenna, permittivities of the media through which the radar wave is traveling, etc. Therefore, the characteristics of the scattered wave received by the SBR antenna for a given frequency depend on the surface and geometric parameters of the radar. Therefore, to apply necessary corrections for scattering of radar waves, the precise knowledge of the instantaneous SBR antenna shape becomes a necessity. However, excitation of flexible dynamics mode by frequent pointing maneuvers makes shape estimation difficult. While a variety of surface models can be employed to model the instantaneous shape, we consider the case that the surface is measured at discrete points using a metrology sensor system, and a smooth least squares approximation is desired. The objective of this section is to evaluate the GLO-MAP methodology developed in

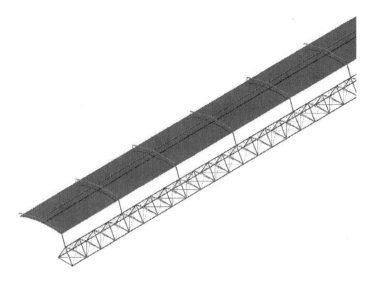

(a) NASTRAN SBR Antenna Model Consisting of Seven Panels

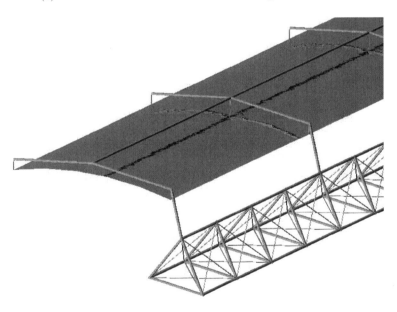

(b) Close-Up of One Panel

FIGURE 5.18
NASTRAN Model of SBR Antenna.

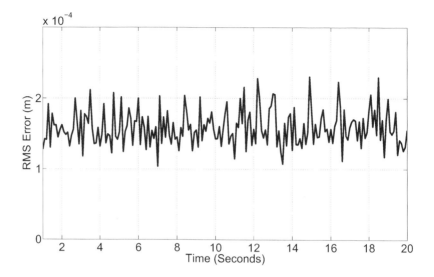

FIGURE 5.19
RMS Approximation Error for SBR Antenna Shape Approximation.

this chapter as a candidate approach to estimate the real-time SBR antenna shape.

For simulation purposes, the SBR antenna dynamics is modeled in NAS-TRAN [73]. The antenna model consists of a total of seven panels as shown in Fig. 5.18. To construct the shape of the antenna, it is assumed that measurements of 50 points are available along a given cross-section with the help of some vision sensor. Further, such 40 cross-section measurements are assumed to be available along the length of the antenna at a particular time with a sampling frequency of 10 Hz. Thus, 2,000 measurements are available every 0.1 seconds. Further, true measurements are corrupted by Gaussian white noise of standard deviation of 1cm. To make the shape estimation problem more interesting, the shape of the antenna is assumed to vary both in spatial position and time. NASTRAN is used to generate mass, $\mathbf{M}$, and stiffness, $\mathbf{K}$, matrices for the antenna structure, and a coordinate transformation matrix, $\mathbf{T}$, to transform the modal coordinates to physical coordinates, i.e., deflections along each axis:

$$\text{Modal Equations: } \mathbf{M}\ddot{\eta} + \mathbf{K}\eta = 0 \qquad (5.96)$$

$$\text{Transformation to Physical Coordinates: } \mathbf{y} = \mathbf{T}\eta \qquad (5.97)$$

where η and $\mathbf{y}$ represent modal and physical coordinates, respectively. First, 10 modes are considered to generate the measurement data. Further, the NASTRAN-generated FEM model with 6,000 degrees of freedom was simu-

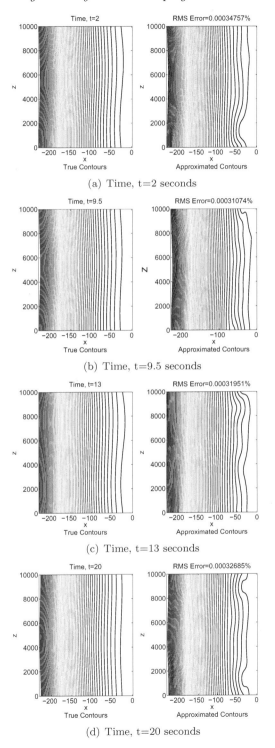

FIGURE 5.20

Space-Based Radar Antenna Simulation Results: True and Approximated Contour Plots.

lated in the MATLAB [74] environment to generate the true measurement data for 20 seconds at 10 Hz frequency.

To approximate the SBR antenna shape at a particular time, the measurement data is modeled using a total of 64 finite element cells, 4 along each of the Cartesian coordinates, X, Y and Z. Now, a continuous approximation of the SBR antenna shape for a particular cell is generated via a least squares procedure as listed in Section 5.4.

The SBR antenna shape at time t for a particular cell is modeled by orthogonal polynomials (given in Table 5.2) of antenna shape at time t_0:

$$\hat{x}(t) = \sum_i \sum_j \sum_k a_{ijk}\phi_i(x(t_0))\phi_j(y(t_0))\phi_k(z(t_0)), \ i+j+k \leq 2 \quad (5.98)$$

$$\hat{y}(t) = \sum_l \sum_m \sum_n a_{lmn}\phi_l(x(t_0))\phi_m(y(t_0))\phi_n(z(t_0)), \ l+m+n \leq 2$$

$$(5.99)$$

It should be noted that x, y and z denote the non-dimensional local cell coordinates defined below:

$$x = 2(X - X_m)/X_{cell} \ y = 2(Y - Y_m)/Y_{cell} \ z = 2(Z - Z_m)/Z_{cell} \quad (5.100)$$

where (X_m, Y_m, Z_m) and $X_{cell} \times Y_{cell} \times Z_{cell}$ represent the centroid and dimensions of a particular cell, respectively. To recursively learn the local approximations at each measurement time, vision sensor measurements are processed sequentially. Initially, all Fourier coefficients are assumed to be zero and the corresponding covariance matrix initialized to 10^6 times the identity matrix.

Further, the true antenna shape is simulated by considering 80 points along each cross-section, and 80 such cross-sections, thus giving rise to a total of $6,400$ test points at each time instant. The first-order weighting function is used to blend adjacent local approximations. Fig. 5.19 shows the plot of RMS approximation error at each time instant, while Fig. 5.20 shows the contour plots for instantaneous antenna shape. From these figures, it is clear that the mean RMS approximation error for X and Y coordinates are even less than half of a percent at all time intervals. Therefore, we can conclude that we are able to learn the SBR antenna shape precisely, even in the presence of measurement errors. Finally, we mention that the simulated antenna shape is just representative and may be a poor approximation of the actual flexible dynamics. In Ref. [75] we use the GLO-MAP algorithm to approximate the flexible body dynamics, instead of modeling just the instantaneous shape measurements.

5.7.4 Porkchop Plot Approximations for Mission to Near-Earth Objects (NEOs)

Near-Earth Objects (NEOs) are asteroids, comets and large meteoroids whose orbits intersect Earth's orbit and which may, therefore, pose a collision danger to Earth. In terms of orbital elements, NEOs are heavenly bodies with their perihelion distance less than 1.3 AU* [76]. Further, NEOs are divided into different groups depending upon many factors, like their orbit size and closest approach to Earth. Out of many near-Earth objects there is an immense interest in visiting near-Earth asteroids due to their size and proximity. There are a total of 701 known near-Earth asteroids which are supposed to make threateningly close approaches (0.05 AU) to Earth [76]. Therefore, information on their structural and chemical compositions is important in order to make intelligent choices in case of an Earth-threatening trajectory. In the last decade, many space missions (NEAR, Deep Impact, STARDUST, MUSES, etc.) were launched to study various near-Earth Objects. For any interplanetary mission design, developing porkchop plots is the most important thing to do. As the name suggests porkchop plots are porkchop-shaped contour plots that display trade space of launch and arrival dates, and help mission designers to find the minimum energy transfer between two planetary objects. Actually, the porkchop plot represents a family of numerical solutions to the two-point boundary value problem known as Lambert's problem [77]. The solution to Lambert's problem gives us a pair of launch and arrival dates that represents a single, unique interplanetary trajectory. So porkchop plots are generated by solving the Lambert problem for the Earth-to-NEO transfer over the range of launch and arrival dates. Later, porkchop plots are used to identify a small region of launch and arrival dates with minimum energy requirements.

There are many algorithms listed in the literature for solving the classic Lambert's problem using the two-body model [77, 78]. However, for a mission to a near-Earth object, one should obtain a solution to the generalized, perturbed Lambert's problem using a general n-body model. This solution is generally based on the use of predetermined reference orbits, by using the two-body model as the first guess and defining the effect of other bodies as perturbation to these reference orbits. Due to high sensitivity of the perturbed solution to the first guess (reference orbits obtained by using the two-body model), one needs to solve and store many reference orbits for different values of position vectors of targeting bodies and time of flight. Further, to obtain the perturbed solution accurately using these reference orbits as a first guess, it is desired to represent these reference orbits as a function of departure date and time of flight. The objective of this section is to apply the GLO-MAP methodology to approximate the porkchop plots for a mission from Earth to asteroid 2003-YN107.

Asteroid 2003-YN107 is a quasi-moon of the Earth in a neighboring solar orbit with time period of 362.264 days. It made its closest approach to Earth (0.03 AU) in June 2005 [76]. To obtain the porkchop plots for this specific mission, we first solve Lambert's problem using the two-body model. The

---

*1 AU $\approx$ 150 million KM.

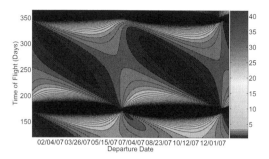

(a) True Porkchop Plot for Departure ΔV_∞

(b) True Porkchop Plot for Arrival ΔV_∞

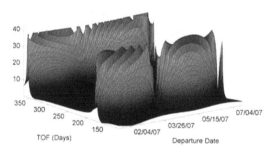

(c) True Surface Plot for Departure ΔV_∞

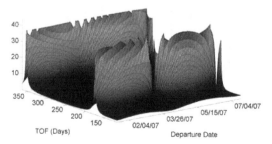

(d) True Surface Plot for Arrival ΔV_∞

FIGURE 5.21

True Departure and Arrival ΔV_∞ Plots for a Mission to the Asteroid 2003-*YN*107.

details of the solution to Lambert's problem are beyond the scope of this book and can be found in Refs. [77, 78].

The Lambert algorithm was used to iteratively solve for the initial launch velocity at a prescribed departure date to reach a target asteroid at a prescribed arrival date. By sweeping the departure and arrival dates, the Lambert algorithm can generate a dense family of solutions. The converged departure and arrival data set can be displayed in porkchop surface plots to be used in mission analysis. Approximating these surfaces enables interpolation of points between the Lambert algorithm points. To approximate the porkchop plots, the measurement data (departure and arrival velocities) are generated by solving Lambert's problem considering a launch window between January 2007 and January 2008 with an interval of 1 day. Further, the Time of Flight (TOF) is varied between 120 to 365 days, giving rise to a total of $365 \times 246 = 89,790$ measurement points for approximation purposes. We should mention that to solve Lambert's problem, the ephemeris data for Earth and asteroid 2003-YN107 are obtained from NASA's near-Earth object program site [76]. Figs. 5.21(a)* and 5.21(b)* show the porkchop plots for departure and arrival velocities, respectively, whereas Figs. 5.21(c)* and 5.21(d)* show the corresponding surface plots. It should be mentioned that sharp peaks in solution make the problem of approximating these solutions extremely difficult. To obtain better approximation accuracy, we use a coarse grid (25×25) in the region where departure and arrival velocity profile is comparatively smooth, i.e., $220 \leq TOF \leq 300$ days, and a relatively finer grid (20×20) in the region where we have sharp peaks, i.e., $TOF < 220$ days and $TOF > 300$ days. Further, to approximate the measurement data in each grid cell, the orthogonal basis functions (listed in Table 5.2) up to quadratic terms in TOF and departure date are used. To compute the value of approximated departure and arrival velocities for a given departure date and TOF, we use a first-order weighting function to blend different local approximations. Figs. 5.22(a)* and 5.22(b)* show the approximated porkchop plots for departure and arrival velocities, respectively, whereas Figs. 5.22(c)* and 5.22(d)* show the corresponding approximated surface plots. Further, Figs. 5.23 and 5.24 show the plots of error contours for departure and arrival velocity for different values of TOF. From these figures, we can conclude that we are able to approximate the porkchop plots for this particular mission with worst case errors less than 0.05 km/sec.

Finally, we also mention that to store the GLO-MAP approximation, one just needs $8(625 + 2, 400) = 11,400$ real numbers, as compared to $3 \times 89,790 = 269,370$ real numbers for the original measurement data. So, the use of the GLO-MAP algorithm reduces the storage space by an order of magnitude in this particular case. In addition, it should be noted that the GLO-MAP

---

*Please see the Appendix for the color versions of these figures.

algorithm allows us to obtain the porkchop plots at any desired resolution without compromising approximation accuracy.

5.8 Summary

We have presented a general methodology for input-output mapping in N dimensions. The method averages overlapping local preliminary approximations whose centroids of validity lie at the vertices of a user-specified, generally non-uniform, N-dimensional grid. The averaging makes use of a special class of weight functions that guarantee a prescribed degree of piecewise continuity and osculation with the preliminary approximations at their centroids of validity. The preliminary approximations can be chosen arbitrarily to take advantage of prior knowledge of a particular problem. Alternatively, the preliminary approximations can be chosen as linear combinations of any complete set of linearly independent basis functions. A particularly attractive choice is shown to be polynomial basis functions that are orthogonal with respect to the weight functions of the averaging process. The result is a new method for orthogonal function local approximation with an associated averaging process giving a global piecewise continuous approximation. A fundamental theoretical result is that if the local approximations are unbiased estimations of the input-output data, then: (*i*) *the final blended approximation is unbiased;* (*ii*) *the variance of the blended approximation is substantially smaller than the variance of any of the averaged local models;* (*iii*) *the mathematical structure of local models can be varied as appropriate to optimally capture the local geometry; and finally,* (*iv*) *the blending weight functions guarantee the global piecewise continuous nature of the approximation.* Further, the new approach is tested on several examples from a variety of disciplines such as continuous function approximation, dynamic system modeling and system identification. The results are of direct utility in addressing the "curse of dimensionality" and frequent redundancy of neural network approximation. The broad generality of the method, together with a number of examples presented, provides a strong basis for optimism for the importance and practical utility of these ideas.

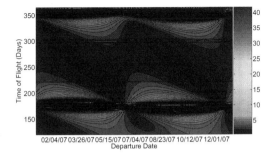

(a) Approximated Porkchop Plot for Departure ΔV_∞

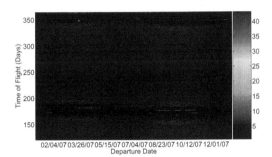

(b) Approximated Porkchop Plot for Arrival ΔV_∞

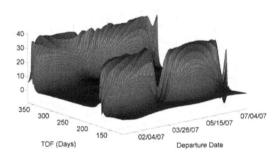

(c) Approximated Surface Plot for Departure ΔV_∞

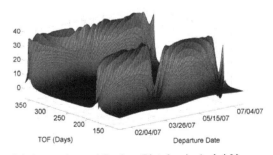

(d) Approximated Surface Plot for Arrival ΔV_∞

FIGURE 5.22
Approximated Departure and Arrival ΔV_∞ Plots for a Mission to the Asteroid 2003-YN107.

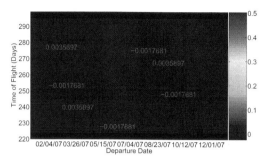

(a) Error Contour Plot for $220 \leq TOF \leq 300$

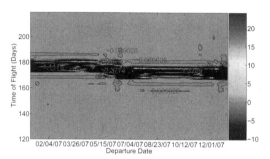

(b) Error Contour Plot for $TOF < 220$

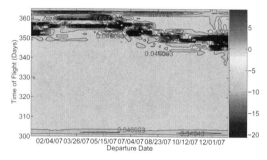

(c) Error Contour Plot for $TOF > 300$

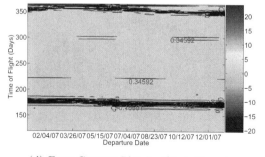

(d) Error Contour Plot for Whole Domain

FIGURE 5.23
Departure ΔV_∞ Approximation Results.

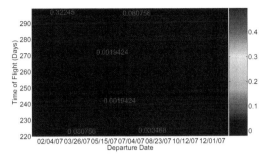

(a) Error Contour Plot for $220 \leq TOF \leq 300$

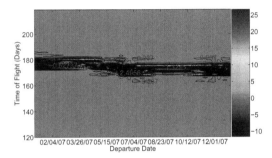

(b) Error Contour Plot for $TOF < 220$

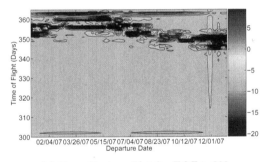

(c) Error Contour Plot for $TOF > 300$

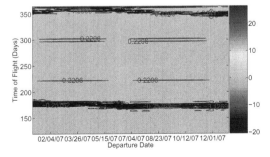

(d) Error Contour Plot for Whole Domain

FIGURE 5.24
Arrival ΔV_∞ Approximation Results.

6

Nonlinear System Identification

This book fills a much-needed gap.

Moses Hadas

6.1　Introduction

System IDentification (SysID) is the term associated with the estimation and validation of mathematical models of physical phenomena from measured input-output data. Usually, SysID implies a general nonlinear dynamical system where the space/time evolution is governed by a poorly known mathematical model consisting of differential equations, integral equations or difference equations. Whereas algebraic "curve fitting" may be concerned with choosing a set of basis functions and estimating their coefficients, SysID frequently begins by fitting a system dynamical model solution (typically a differential, difference or integral equation model).

SysID is a most fundamental step in virtually all disciplines of science and engineering. Dynamical system models are used for the design and analysis of complex technical systems. Various classical and modern controller design techniques, such as the Linear Quadratic Regulator (LQR) and Lyapunov controller, require a dynamical model relating the control variables to the system output. Of course, dynamical models can frequently be constructed from first principles, but it is also of vital importance that they can be approximated or refined directly from measurements.

In the last five decades, system identification theory has evolved into a powerful scientific tool of wide applicability. However, the most mature part of the theory deals with linear systems using well-established techniques of linear algebra and the theory of ordinary differential or difference equations. In contrast, nonlinear system identification problems are still treated mostly on a system-by-system basis. In this chapter, our main interest is to introduce some general nonlinear system identification techniques that can be applied to represent state or output perturbation for a large class of systems for which linear approximation is not adequate. We also consider problems with high state space dimension, for example, models for the dynamics of large flexible space structures.

This chapter is written with four main objectives. The first and most important objective is to present a novel robust nonlinear system identification method using the GLO-MAP system representation introduced in Chapter 5. The second objective is to present adaptive learning algorithms to adjust in real time the parameters of the GLO-MAP model. The learning algorithm introduced in this chapter is inspired by recent developments in adaptive control [79, 80]. The third objective is to compare the nonlinear system identification algorithm with some existing identification algorithms like the Eigensystem Realization Algorithm (ERA) [81], considering applications involving modeling of large flexible space structures. The fourth and final objective is to set down a theoretical framework, including all assumptions, that guarantees the stability of the new algorithm introduced. Because these theoretical results have very few companion results in the existing nonlinear system identification literature, special care is taken to clearly state all the assumptions and develop theoretical conditions for stability.

The structure of this chapter is as follows: First the system identification problem is introduced followed by a brief review of some existing system identification algorithms. We give special attention to the Eigensystem Realization Algorithm (ERA) because of its broad utility and numerical robustness for linear and near linear systems. Then, two different robust system identification algorithms are introduced using the GLO-MAP modeling approach [62, 65], discussed in Chapter 5. Finally, the new algorithms are validated and compared by simulating test cases concerned with large space structure applications.

6.2   Problem Statement and Background

Let us consider a nonlinear system described by the following differential and algebraic equations:

$$\dot{\mathbf{x}}(t) = f\left(\mathbf{x}(t), \mathbf{u}(t)\right) \tag{6.1}$$

$$\mathbf{y}(t) = g\left(\mathbf{x}(t), \mathbf{u}(t)\right) \tag{6.2}$$

where $\mathbf{x} \in \mathcal{R}^n$ and $\mathbf{u} \in \mathcal{R}^p$ represent state and control vectors, respectively, and $\mathbf{y} \in \mathcal{R}^m$ represents a vector of system outputs at time t. The discrete analog of this system is the following nonlinear difference equations:

$$\mathbf{x}_k = f_d\left(\mathbf{x}_{k-1}, \mathbf{u}_{k-1}\right) \tag{6.3}$$

$$\mathbf{y}_k = g_d\left(\mathbf{x}_k, \mathbf{u}_k\right) \tag{6.4}$$

Now, if the functions, $f(.)$, $g(.)$, $f_d(.)$, and $g_d(.)$ are unknown, then the system identification problem can be formally stated as follows:

Definition of the System Identification Problem. *Identify a mathematical model which, when subject to the actual input vector,* $\mathbf{u}$, *an output estimate* $\hat{\mathbf{y}}$ *is produced which approximates the actual system output,* $\mathbf{y}$, *such that*

$$\|\mathbf{y} - \hat{\mathbf{y}}\| \leq \epsilon \tag{6.5}$$

Here, $\|.\| : \mathcal{R}^m \rightarrow \mathcal{R}$ represents a suitable norm on the system output space, $\mathbb{Y}$, and ϵ dictates the desired accuracy of the system identification problem. In other words, the system identification problem corresponds to finding a model whose outputs are as close as desired to the true system outputs when the same input is applied to both. Therefore, the system identification problem can also be regarded as the identification of a continuous map from system input space to system output space [82]. Consequently the problem of approximating a continuous functional arises in the system identification problem. The output at any given time is considered a function of the input signal, which is a function of time. Implicitly, we hope the input-output data approximated is sufficiently rich that the model will be accurate over a wide class of inputs, and useful for other purposes such as controlling the system.

Various system identification algorithms are described in the literature for input-output mapping [36, 82–87]. The main computational tool employed by these algorithms is the process of Least Squares Estimation (LSE), frequently implemented using the Singular Value Decomposition (SVD). The LSE method along with SVD results in numerical robustness under very weak assumptions on the persistency of excitation of the inputs. In the past few decades, Artificial Neural Networks (ANNs) have emerged as a powerful set of tools in the areas of pattern classification, time series analysis, signal processing, dynamical system modeling, and control. The emergence of ANNs can be attributed to the fact that these network models are frequently able to learn behavior when traditional modeling is very difficult to generalize. However, the optimal number of hidden units, perceptrons, depends upon many factors, like the ability of the chosen basis functions to approximate the given system's behavior, the number of data points, the signal to noise ratio, the complexity of the learning algorithms, etc. Narendra et al. [82, 84] have proposed different models that utilize two-layered neural networks with sigmoid functions as activation functions for system identification. In those papers, the output signal at any time is considered a function of finitely many samples of the input and output signals. The different ANN parameters are estimated using a back-propagation algorithm [15]. A key issue arises because if one fixes the architecture and activation functions, a given ANN's ability to approximate a given system's behavior can be deduced only after the learning process is over. Adaptation of the network architecture, not simply adjusting weights in a fixed architecture, has emerged as the key to convergence reliability and accuracy.

In Chapter 3, an adaptive RBFN architecture, making use of the Directed Connectivity Graph (DCG) algorithm, is introduced and used for many example system identification problems. It has been shown that the adaptive nature of the network improves significantly the performance of the algorithm in terms of accuracy and number of free network parameters, in comparison to many traditional algorithms. However, like traditional ANN algorithms, the DCG algorithm also treats the system identification problem as the identification of a global continuous map from system input space to system output space. As a consequence of this, the performance of these algorithms decreases drastically as the dimension of the system output vector increases. To make this point more clear, consider a problem of active control of a flexible space structure. To derive a control law, a model of the system dynamics from the control variable, $\mathbf{u}(t)$, to the system output, $\mathbf{y}(t)$, is desired. Generally, the system output vector consists of surface distortion measurements at various spatial points, $O(10^3)$, which are measured by sensors like strain gauges, stereo vision systems, LIDAR, etc. Therefore, if one seeks a dynamic continuous map between the system output and input vectors then the dimension of such a map can be as large as the number of measurements, i.e., $O(10^3)$. A conventional finite element model, before model reduction, will have in excess of $O(10^5)$ spatial degrees of freedom. However, the dimension of the hidden states corresponding to the true system corresponds to the number of dynamic structural modes of interest, which are typically on the order of 10 to 30. So, a system identification algorithm is desired that can approximate the system output well, while keeping the dimension of the dynamic map as low as possible. To deal with this challenge, various model reduction techniques are often adopted [88] for approximating high order dynamic models by simpler, lower order models. The most popular method for model reduction is Proper Orthogonal Decomposition (POD) [88], also known as the Principal Component Analysis (PCA). In model reduction, one would like to preserve properties of the original model, such as stability and physically important dynamical mode shapes. However, as POD uses second-order statistics for model reduction, it sometimes de-emphasizes infrequent events which can be dynamically very important.

6.3   Novel System Identification Algorithm

In the previous section, issues concerning the inability of various system identification algorithms to handle the "curse of dimensionality" are discussed. In this section, two novel nonlinear system identification algorithms are introduced which make use of the classical Eigensystem Realization Algorithm (ERA) [81] and the recently developed Global-Local Orthogonal Polynomial

Mapping (GLO-MAP) [62] network to deal with the issues of nonlinearity and high-dimensioned output vectors in an efficient manner. These novel algorithms not only have the approximation ability of the ANNs but also have the model reduction ability of algorithms like POD.

The basic idea of both of the new algorithms is to split the identification process into two steps: linear system identification followed by the nonlinear system identification process. The linear system identification process not only helps in designing an estimator to estimate hidden dynamic states from the measurement data, but also gives an approximate dimension for the reduced order dynamical model for the hidden state vector. It implicitly defines a transformed state space that is physically motivated to capture the best linear representation of the the system input-output behavior. We elect to retain this linear transformation as the starting point for a perturbation approach to account for the nonlinear departure from this best linear model. The use of the best fitting linear system to establish a desired order state space model is an attractive feature of the new algorithms which also helps in dealing with the "curse of dimensionality."

The first step of both algorithms is to identify a linear dynamical system from the time history of input-output data. Referring to Figs. 6.1 and 6.2, let the best linear model ("realization") be written as:

$$\dot{\mathbf{x}}_l = \mathbf{A}_l \mathbf{x}_l + \mathbf{B}_l \mathbf{u} \tag{6.6}$$

$$\mathbf{y}_l = \mathbf{C}_l \mathbf{x}_l + \mathbf{D}_l \mathbf{u} \tag{6.7}$$

Here, $\mathbf{x}_l \in \mathcal{R}^n$ is a hidden state vector corresponding to the best linear approximation of given input-output data. While $\mathbf{A}_l$, $\mathbf{B}_l$, $\mathbf{C}_l$, $\mathbf{D}_l$ are not unique, the underlying input-output map is, and the balanced $\mathbf{A}_l$, $\mathbf{B}_l$, $\mathbf{C}_l$, $\mathbf{D}_l$ realization from the ERA can be robustly computed, including the dimension n of the state space. The accuracy of the output vector, $\mathbf{y}_l$, in approximating the true system measured output data, $\mathbf{y}$, depends upon the nonlinearities involved. We mention that the ERA methodology for differentiation of "noise modes" due to random measurement from the actual system is reduced order dynamics that has matured to a practical stage [85]. In the event that the measurement output error exceeds some tolerance, i.e., $\|\mathbf{y} - \mathbf{y}_l\| > \epsilon$, the linear model must be judged invalid and the need to generalize Eqs. (6.6) and (6.7) arises.

Two different sequential algorithms are introduced below to apply the correction to the best identified linear system.

1. In the first (*state model perturbation*) approach, the linear state dynamical model of Eq. (6.6) is perturbed by a nonlinear term to learn the difference between the linear propagated state vector, $\mathbf{x}_l$, and the best estimate of state vector, $\mathbf{x}_b$.

$$\dot{\hat{\mathbf{x}}} = \mathbf{A}_l \hat{\mathbf{x}} + \hat{\mathbf{B}} \mathbf{u} + \mathbf{g}(\hat{\mathbf{x}}) \tag{6.8}$$

$$\hat{\mathbf{y}} = \mathbf{C}_l \hat{\mathbf{x}} + \mathbf{D}_l \mathbf{u} \tag{6.9}$$

Here, $\mathbf{g}(\hat{\mathbf{x}})$ is a vector of unknown nonlinearities which can be learned by conventional ANN methods. Note we have made $\mathbf{g}$ depend on $\hat{\mathbf{x}}$, the same reduced order state vector that resulted from the best fitting linear system. To find the best estimates of the hidden dynamic state vector, $\hat{\mathbf{x}}_b$, from the given system output data, $\mathbf{y}$, an efficient estimator such as an algebraic Kalman filter can be designed using the measurement model of Eq. (6.9). The main steps of this algorithm are illustrated in Fig. 6.1 and we denote this approach by short form SysID 1. For the purpose of this chapter, $\mathbf{g}(.)$ represents the system nonlinearities not captured by the linear model and modeled by using the GLO-MAP process of Chapter 5. Note that the performance of this approach depends upon the dimensionality of the hidden state vector, $\mathbf{x}$, and the frequency at which hidden state vector estimates can be obtained.

2. In the second (*output model perturbation*) approach, we propose the design of a Kalman filter using the best known linear model followed by the nonlinear transformation of the estimated output data to compensate for their deviation from true output data, $\mathbf{y}$.

$$\hat{\mathbf{y}}_l = \mathbf{C}_l \hat{\mathbf{x}}_l + \mathbf{D}_l \mathbf{u} \qquad (6.10)$$

$$\hat{\mathbf{y}} = \hat{\mathbf{y}}_l + \mathbf{h}(\mathbf{x}_p) \qquad (6.11)$$

Here $\hat{\mathbf{x}}_l$ represents the state output of the linear Kalman filter. Note the design of the Kalman filter helps us in reducing the propagation error arising due to system nonlinearities if output data, $\mathbf{y}$, is available at a reasonable frequency. Further, $\mathbf{h}(.)$ represents the system nonlinearities not captured by the linear model and can be modeled by traditional ANN algorithms. To keep the dimensionality of the nonlinear transformation $\mathbf{h}(.)$ low, we introduce a new variable, $\mathbf{x}_p$. The dummy variable, $\mathbf{x}_p$, can be regarded as a physical variable associated with the problem at hand and is not necessarily be the same as hidden state vector $\mathbf{x}_l$. For example, in the case of the modeling of a flexible space structure, the system output vector consists of surface distortion measurements at various spatial points; therefore, the dummy variable $\mathbf{x}_p$ can consist of Cartesian coordinates (x, y, z). In other words, the surface distortions can be modeled as a function of undeformed Cartesian coordinates. We find, in many problems, a physically motivated low-dimensioned $\mathbf{x}_p$ can be chosen which leads to an accurate nonlinear input-output map. The overall architecture of this algorithm is illustrated in Fig. 6.2. For the purpose of this chapter, $\mathbf{h}(.)$ is modeled by using the GLO-MAP process of Chapter 5, and short form SysID 2 is used to describe this approach. Note, the performance of this approach depends upon the number of measurement points available to learn $\mathbf{h}(.)$ and the frequency at which measurement data are available.

We mention that, ideally, a combination of both the *state model* and *output model* perturbation approaches is desired for efficient and accurate modeling of the dynamical systems. However, the use of the state model perturbation approach is recommended if the dimensionality of the hidden state vector $\mathbf{x}_l$ is low because as the the dimensionality of the hidden state vector $\mathbf{x}_l$ increases the number of terms required to model $\mathbf{g}(.)$ increases exponentially. Further, the use of the output model perturbation approach is highly desirable for the modeling of a flexible space structure when the participating modes are large in number, i.e., $O(10\text{-}30)$. We mention that the system-identification problem as stated in this section does not deal with the issue of uniqueness of the mathematical model. This issue cannot be dealt with theoretically in general for nonlinear systems, but this criticism is not unique to our approach. We find that this theoretical deficiency is usually not an obstacle to practical progress. We note that the main practical objective of the system identification process is the generation of a workable mathematical model with accuracy sufficient for a particular application. Finally, the convergence of both the algorithms is an important issue and will be studied later in this chapter after discussing each step in detail.

6.3.1   Linear System Identification

As discussed in the previous section, linear system identification plays an important role in the success of both of the nonlinear system identification algorithms. The linear system identification process not only helps in designing an estimator to estimate hidden dynamic states from sensor noise-corrupted measurement data, but also gives a desired order dynamical model for the hidden state vector. A large class of linear system identification methods [85–87] are addressed in the literature to estimate hidden state variables along with the dynamical model from given input-output data. However, we believe the ERA and the Observer/Kalman filter IDentification (OKID) algorithms [81, 87] are among the most popular for dynamic system modeling, and have been successfully used in various system identification problems for structural analysis. The first realization, i.e., ERA, has been found to be particularly robust and useful for structural dynamic systems where one is interested in identifying the dominate modes, eigenvalues, and modal shapes; these may be identified directly by this approach. The ability to identify only the modes actually participating in the measured behavior of the system helps in dramatically reducing the order of the system and thus implicitly dealing with the "curse of dimensionality." The ERA algorithm is recommended for the linear system identification module in both of the algorithms (see Figs. 6.1 and 6.2); however, any other linear system identification algorithm can be used instead of the ERA. In this section, the main steps of the ERA algorithm are briefly discussed and more details can be found in Ref. [81].

1. The first step of the ERA method is to form the Hankel matrix from the measurement outputs, $\mathbf{Y}(t_k)$, according to the following pattern of

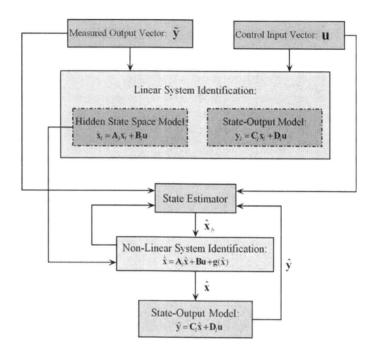

FIGURE 6.1
Overall Architecture of the State Model Perturbation Approach to the System
Identification Algorithm.

time-shifting:

$$
\mathbf{H}_{rs}(k-1) = \begin{bmatrix} \mathbf{Y}(k) & \mathbf{Y}(k+t_1) & \cdots & \mathbf{Y}(k+t_{s-1}) \\ \mathbf{Y}(j_1+k) & \mathbf{Y}(j_1+k+t_1) & \cdots & \mathbf{Y}(j_1+k+t_{s-1}) \\ \vdots & \vdots & & \vdots \\ \mathbf{Y}(j_r+k) & \mathbf{Y}(j_r+k+t_1) & \cdots & \mathbf{Y}(j_r+k+t_{s-1}) \end{bmatrix} \quad (6.12)
$$

Further, it can be easily shown that the Hankel matrix expression generalizes to the following factored expression:

$$
\mathbf{H}_{rs}(k) = \mathbf{V}_r \mathbf{A}^k \mathbf{W}_s \quad (6.13)
$$

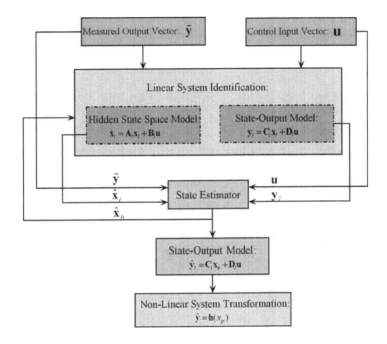

FIGURE 6.2
Overall Architecture of the Output Model Perturbation Approach to the System Identification Algorithm.

where $\mathbf{V}_r$ is the observability matrix given by the following expression:

$$\mathbf{V}_r = \begin{bmatrix} \mathbf{C} \\ \mathbf{C}\mathbf{A}^{j1} \\ \vdots \\ \mathbf{C}\mathbf{A}^{j_r-1} \end{bmatrix} \tag{6.14}$$

whereas $\mathbf{W}_s$ is a controllability or disturbability matrix, given by the following equations depending upon the control input:

Impulse Response (IR): $\mathbf{W}_s = \begin{bmatrix} \mathbf{B} & \mathbf{A}^{t1}\mathbf{B} & \cdots & \mathbf{A}^{t_s-1}\mathbf{B} \end{bmatrix}$ (6.15)

Initial State Response (ISR): $\mathbf{W}_s = \begin{bmatrix} \mathbf{B} & \mathbf{A}^{t1}\mathbf{X}_0 & \cdots & \mathbf{A}^{t_s-1}\mathbf{X}_0 \end{bmatrix}$ (6.16)

Of course, Eq. (6.13) is the "forward problem" representation for given $\mathbf{A}, \mathbf{B}, \mathbf{C}, \mathbf{D}$. The desired representation is the solution of the "inverse problem": *find* $\mathbf{A}, \mathbf{B}, \mathbf{C}, \mathbf{D}$ *from the Hankel matrix of Eq. (6.12).*

2. In the second step, a matrix $\mathbf{H}^\sharp$ is desired such that following is true:

$$\mathbf{W}_s\mathbf{H}^\sharp\mathbf{V}_r = \mathbf{I}_n \qquad (6.17)$$

The singular value decomposition is truncated by retaining the non-zero singular values. In the presence of measurement noise, Juang et al. [85] have developed modal coherence criteria to truncate "noise modes" associated with noise in the measured Hankel matrix.

A general solution for $\mathbf{H}^\sharp$ is found by the Singular Value Decomposition (SVD) of $\mathbf{H}_{rs}(0) = \mathbf{P}\mathbf{D}\mathbf{Q}^T$:

$$\mathbf{H}^\sharp = \mathbf{Q}\mathbf{D}^{-1}\mathbf{P}^T \qquad (6.18)$$

3. Finally, after some algebraic manipulations, the following relationship for $\mathbf{Y}(k+1)$ is obtained:

$$\mathbf{Y}(k+1) = \underbrace{\mathbf{E}_p^T\mathbf{P}\mathbf{D}^{1/2}}\left[\underbrace{\mathbf{D}^{-1/2}\mathbf{P}^T\mathbf{H}_{rs}(1)\mathbf{Q}\mathbf{D}^{-1/2}}\right]^k \underbrace{\mathbf{D}^{1/2}\mathbf{Q}^T\mathbf{E}_m} \quad (6.19)$$

where $\mathbf{E}_k^T = \begin{bmatrix} \mathbf{I}_k & \mathbf{O}_k & \cdots & \mathbf{O}_k \end{bmatrix}$. Comparing this relationship with Eqs. (6.15) and (6.16), the following realizations for matrices $\mathbf{A}$, $\mathbf{B}$ and $\mathbf{C}$ are obtained for the case of Impulse Response (IR) and Initial State Response (ISR), respectively:

IR: $\mathbf{A} = \mathbf{D}^{-1/2}\mathbf{P}^T\mathbf{H}_{rs}(1)\mathbf{Q}\mathbf{D}^{-1/2}$, $\mathbf{B} = \mathbf{D}^{1/2}\mathbf{Q}^T\mathbf{E}_m$, $\mathbf{C} = \mathbf{E}_p^T\mathbf{P}\mathbf{D}^{1/2}$

ISR: $\mathbf{A} = \mathbf{D}^{-1/2}\mathbf{P}^T\mathbf{H}_{rs}(1)\mathbf{Q}\mathbf{D}^{-1/2}$, $\mathbf{X}_0 = \mathbf{D}^{1/2}\mathbf{Q}^T\mathbf{E}_m$, $\mathbf{C} = \mathbf{E}_p^T\mathbf{P}\mathbf{D}^{1/2}$

Now, let the estimated state matrix, $\mathbf{A}$, be of order n and have a complete set of linearly independent eigenvectors $(\psi_1, \psi_2, \cdots, \psi_n)$ with corresponding eigenvalues $(\lambda_1, \lambda_2, \cdots, \lambda_n)$ which are not necessarily distinct. Define $\mathbf{\Lambda}$ as the diagonal matrix of eigenvalues, and $\mathbf{\Psi}$ as the matrix whose columns are the eigenvectors. Then, the minimum state-realization can be transformed to a minimum modal-realization. The diagonal matrix $\mathbf{\Lambda}$ contains the information about modal damping rates and damped natural frequencies, which are simply the real and imaginary parts of the eigenvalues, after transformation from discrete to continuous-time domain via $\mathbf{\Lambda}_c = ln(\mathbf{\Lambda})/\delta t$. The columns of the matrix $\mathbf{\Psi}^{-1}\mathbf{B}$ define the initial modal amplitudes, or information that indicates how effective a particular input is at exciting each mode. The columns of matrix $\mathbf{C}$ define the transformation from modal coordinates to the physical coordinates, i.e., system outputs. Finally, the continuous time representation $(\mathbf{A}_l, \mathbf{B}_l, \mathbf{C}_l, \mathbf{D}_l)$ analogous to discrete-time realization $(\mathbf{A}, \mathbf{B}, \mathbf{C}, \mathbf{D})$ can be obtained easily by assuming a zero-order hold (ZOH) on the inputs.

6.3.2 State Variable Estimation

The hidden state variable estimation is an important step in both of the system identification algorithms given the input-output data and the best learned linear system. Among various estimation algorithms listed in the literature, the Kalman filter [33] is the most widely used for dynamical state identification.

Kalman filtering is a modern (since 1960) development in the field of estimation [2, 6], although it has its roots as far back as in Gauss' work in the 1800s. The only qualitative difference between the Kalman filter and the sequential version of the Gaussian least squares is that the Kalman filter uses a dynamical model of the plant to propagate the state estimates and the corresponding error covariance matrix between two sets of measurements. In this section, the Kalman filter algorithm is described to find the best estimate of the hidden state variable, $\hat{\mathbf{x}}_b$, for the proposed nonlinear system identification algorithms.

The various steps involved in the estimation of hidden state variables using the Kalman filter are listed as:

1. **Propagation**: This step involves the propagation of the estimated hidden state variable, $\hat{\mathbf{x}}$, and its corresponding state error covariance matrix, $\mathbf{P}_x$, using the best-known dynamical system and the corresponding Ricatti equation:

$$\dot{\hat{\mathbf{x}}} = \mathbf{A}\hat{\mathbf{x}} + \mathbf{B}\mathbf{u} + \mathbf{G}\mathbf{w} \qquad (6.20)$$
$$\dot{\mathbf{P}}_x = \mathbf{P}_x\mathbf{A} + \mathbf{A}^T\mathbf{P}_x + \mathbf{G}\mathbf{Q}\mathbf{G} \qquad (6.21)$$

Here $\mathbf{w}$ represents the process noise vector modeled as Gaussian white noise with known covariance matrix $\mathbf{Q}$. It should be mentioned that Eq. (6.20) represents the best-known differential equation for the evolution of hidden state variable $\mathbf{x}$. Eqs. (6.6) and (6.8) are used for hidden state propagation in the system identification approaches SysID 2 and SysID 1, respectively.

2. **Update**: Given the measurement vector, $\tilde{\mathbf{y}}$, at any time, t, the algebraic relationship between the state vector, $\mathbf{x}$, and the system output vector, $\mathbf{y}$, is used to update the propagated estimates of the unknown hidden state vector, $\mathbf{x}^-$, and the corresponding error covariance matrix, $\mathbf{P}_x^-$:

$$\tilde{\mathbf{y}} = \mathbf{H}\mathbf{x} + \mathbf{D}\mathbf{u} + \boldsymbol{\nu} \qquad (6.22)$$

where $\boldsymbol{\nu}$ denotes the measurement noise vector modeled as Gaussian white noise with known covariance matrix $\mathbf{R}$. The following expression can be derived for the state vector estimates using the least squares criteria described in Ref. [2]:

$$\mathbf{K} = \mathbf{P}_x^- \mathbf{H}^T \left(\mathbf{H} \mathbf{P}_x^- \mathbf{H}^T + \mathbf{R} \right)^{-1}$$
$$\hat{\mathbf{x}} = \hat{\mathbf{x}}^- + \mathbf{K} \left(\tilde{\mathbf{y}} - \hat{\mathbf{y}} \right)$$
$$\mathbf{P}_x = \left(\mathbf{I} - \mathbf{KH} \right) \mathbf{P}_x^-$$

The main assumption in using the above expression is *complete observability* of the state vector, $\mathbf{x}$, from Eq. (6.22). In other words, to estimate $\mathbf{x}$ using Eq. (6.22), the matrix $\mathbf{H}$ should have its rank equal to the dimension of $\mathbf{x}$, i.e., n. It should be mentioned that Eq. (6.7) is used as a counterpart to Eq. (6.22) for both of the system identification algorithms described in Section 6.3.

6.4   Nonlinear System Identification Algorithm

In previous sections, we have introduced two nonlinear system identification algorithms. However, we have not discussed, in detail, the procedure to learn the nonlinear terms in both of the algorithms. In this section, first, an adaptive learning algorithm is described to update the linear dynamic model using the hidden state estimates as measurements in the case of SysID 1, followed by the description of the learning algorithm for SysID 2.

6.4.1   Learning Algorithm for State Model Perturbation Approach (SysID 1)

The learning algorithm for SysID 1 is based upon the recently developed GLO-MAP network and uses Lyapunov's stability theorem [89] to determine the update laws for different parameters of the GLO-MAP network.

Consider the perturbed linear dynamic model, where $\mathbf{g}(\mathbf{x})$ is a vector of nonlinear terms introduced as an additive perturbation to a best linear model of the system state differential equation:

$$\dot{\mathbf{x}} = \mathbf{A}_l \mathbf{x} + \mathbf{B}\mathbf{u} + \mathbf{g}(\mathbf{x}) \qquad (6.23)$$
$$\mathbf{y} = \mathbf{C}_l \mathbf{x} + \mathbf{D}_l \mathbf{u} \qquad (6.24)$$

Here, $\mathbf{A}_l \in \mathcal{R}^{n \times n}$ is a known Hurwitz matrix, and $\mathbf{B} \in \mathcal{R}^{n \times p}$ is a control effectiveness matrix. Thus, $\mathbf{g}(\mathbf{x})$ represents the unknown model nonlinearities, and parameterization of these model correction terms is the first step in the state perturbation approach to SysID. It should be noted that the matrix $\mathbf{A}_l$ is chosen in such a way that it captures the modal frequencies of interest, and can be obtained by the ERA or any other suitable linear system identification algorithm, as described in Section 6.3.

The time history estimates of the hidden state vector, $\mathbf{x}$, can be obtained by using the procedure described in Section 6.3, so the system identification problem can be redefined as:

System Identification Problem. *Given the time history estimates of the state vector, $\mathbf{x}(t)$, and control variable, $\mathbf{u}(t)$, find estimates of the unknown nonlinearity vector, $\mathbf{g}(.)$, and control effectiveness matrix, $\mathbf{B}$.*

Further, if $\mathbf{g}(.)$ is assumed to be a continuous function in $\mathbf{x}$, then according to the Weierstrass approximation theorem [7, 8], $\mathbf{g}(.)$ can be approximated arbitrarily close by any set of complete functions, including a polynomial series:

$$\mathbf{g}(\mathbf{x}) = \mathbf{C}^T \mathbf{\Phi}(\mathbf{x}) + \epsilon \qquad (6.25)$$

where $\mathbf{\Phi}(.)$ is an infinite dimensional vector of polynomial functions, $\mathbf{C}$ is a matrix of Fourier coefficients corresponding to polynomial functions, and ϵ denotes the residual approximation error. Usually, $\mathbf{\Phi}(.)$ can be chosen as a finite dimensional vector of orthogonal polynomials. Therefore, $\mathbf{C} \in \mathcal{R}^{N \times n}$ is a matrix of Fourier coefficients corresponding to these orthogonal polynomial functions. Now, substituting Eq. (7.16) in Eq. (6.23) yields:

$$\dot{\mathbf{x}}(t) = \mathbf{A}_l \mathbf{x}(t) + \mathbf{B}\mathbf{u}(t) + \mathbf{C}^T \mathbf{\Phi}(\mathbf{x}) + \epsilon \qquad (6.26)$$

But the Fourier coefficient matrix $\mathbf{C}$ is unknown so we write an estimate equation

$$\dot{\hat{\mathbf{x}}}(t) = \mathbf{A}_l \hat{\mathbf{x}}(t) + \hat{\mathbf{B}}\mathbf{u}(t) + \hat{\mathbf{C}}^T \mathbf{\Phi}(\mathbf{x}) \qquad (6.27)$$

Let us define $\mathbf{e}(t) = \mathbf{x}(t) - \hat{\mathbf{x}}(t)$, which leads to the following expression:

$$\dot{\mathbf{e}}(t) = \mathbf{A}_l \mathbf{e} + (\mathbf{B} - \hat{\mathbf{B}})\mathbf{u}(t) + (\mathbf{C} - \hat{\mathbf{C}})^T \mathbf{\Phi}(\mathbf{x}) + \epsilon \qquad (6.28)$$

$$= \mathbf{A}_l \mathbf{e} + \tilde{\mathbf{B}}\mathbf{u}(t) + \tilde{\mathbf{C}}^T \mathbf{\Phi}(\mathbf{x}) + \epsilon \qquad (6.29)$$

To find adaptation laws for the unknown parameters, we consider the following Lyapunov function:

$$V = \frac{1}{2}\mathbf{e}^T \mathbf{P}\mathbf{e} + \frac{1}{2}Tr(\tilde{\mathbf{B}}\Gamma_1 \tilde{\mathbf{B}}^T) + \frac{1}{2}Tr(\tilde{\mathbf{C}}^T \Gamma_2 \tilde{\mathbf{C}}) \qquad (6.30)$$

where $\mathbf{P}$ is a positive definite symmetric matrix. Now taking the time derivative of V leads to the following equation:

$$\dot{V} = \frac{1}{2}\mathbf{e}^T \underbrace{\left(\mathbf{P}\mathbf{A}_l + \mathbf{A}_l^T \mathbf{P}\right)}_{-\mathbf{Q}} \mathbf{e} + \mathbf{e}^T \mathbf{P}\left(\tilde{\mathbf{B}}\mathbf{u}(t) + \tilde{\mathbf{C}}^T \mathbf{\Phi}(\mathbf{x}) + \epsilon\right) + Tr\left(\tilde{\mathbf{B}}\Gamma_1 \dot{\tilde{\mathbf{B}}}^T\right)$$

$$+ Tr\left(\tilde{\mathbf{C}}^T \Gamma_2 \dot{\tilde{\mathbf{C}}}\right) \qquad (6.31)$$

$$= -\frac{1}{2}\mathbf{e}^T \mathbf{Q}\mathbf{e} + Tr\left(\tilde{\mathbf{B}}\left[\Gamma_1 \dot{\tilde{\mathbf{B}}}^T + \mathbf{u}\mathbf{e}^T \mathbf{P}\right]\right) + Tr\left(\tilde{\mathbf{C}}^T \left[\Gamma_2 \dot{\tilde{\mathbf{C}}} + \mathbf{\Phi}(\mathbf{x})\mathbf{e}^T \mathbf{P}\right]\right)$$

$$+ \mathbf{e}^T \mathbf{P}\epsilon \qquad (6.32)$$

Note here $\mathbf{Q} \in \mathcal{R}^{n \times n}$ is a positive definite matrix which satisfies the following algebraic Ricatti equation:

$$\mathbf{P}\mathbf{A}_l + \mathbf{A}_l^T \mathbf{P} = -\mathbf{Q} \tag{6.33}$$

Now, if the following adaptation laws are chosen for $\tilde{\mathbf{B}}$ and $\tilde{\mathbf{C}}$,

$$\dot{\tilde{\mathbf{B}}}^T = -\dot{\hat{\mathbf{B}}}^T = -\Gamma_1^{-1}\mathbf{u}\mathbf{e}^T\mathbf{P} \tag{6.34}$$

$$\dot{\tilde{\mathbf{C}}} = -\dot{\hat{\mathbf{C}}} = -\Gamma_2^{-1}\mathbf{\Phi}(\mathbf{x})\mathbf{e}^T\mathbf{P} \tag{6.35}$$

then $\dot{V}$ reduces to:

$$\dot{V} = -\frac{1}{2}\mathbf{e}^T\mathbf{Q}\mathbf{e} + \epsilon^T\mathbf{P}\mathbf{e} \tag{6.36}$$

$$\Rightarrow \dot{V} \leq -\frac{1}{2}|\lambda_{min}(\mathbf{Q})|\|\mathbf{e}\|^2 + \|\epsilon\|\|\mathbf{P}\|\|\mathbf{e}\| \tag{6.37}$$

Note:

- When $\epsilon = 0$, i.e., there are no approximation errors, we have the following expression for $\dot{V}$:

$$\dot{V} = -\frac{1}{2}\mathbf{e}^T\mathbf{Q}\mathbf{e} \leq 0 \tag{6.38}$$

  Now the convergence of tracking residual, $\mathbf{e}$, follows from the assumption that $\mathbf{e} \in L_\infty$, i.e., both $\mathbf{x}$ and $\hat{\mathbf{x}}$ are bounded signals. Further, from the integral of $\dot{V}$, it can be easily shown that $\mathbf{e} \in L_2 \cap L_\infty$, and therefore, from Barbalat's Lemma [79] $\mathbf{e} \to 0$ as $t \to \infty$, which in turn leads to $\tilde{\mathbf{B}} \to 0$ and $\tilde{\mathbf{C}} \to 0$, based on Eqs. (6.34) and (6.35). We mention that although $\tilde{\mathbf{B}}$ and $\tilde{\mathbf{C}}$ approach 0, their convergence to corresponding true values is not guaranteed without the satisfaction of the *persistence of excitation* condition [79].

- For a given level of the tracking errors, $\mathbf{e}$, we can only conclude bounded stability as long as the approximation error ϵ satisfies the following bound:

$$\|\epsilon\| \leq \frac{|\lambda_{min}(\mathbf{Q})|\|\mathbf{e}\|}{2\|\mathbf{P}\|} = \epsilon_{ub} \tag{6.39}$$

  The above inequality gives us an upper bound on the approximation error, ϵ, to guarantee the bounded stability of the system identification error, $\mathbf{e}$. Recall that, in Chapter 5, we provide a lower bound on approximation errors which can be used to find a conservative estimate on the number of polynomial basis functions required to approximate $\mathbf{g}(.)$ so that Eq. (6.39) is satisfied. However, if the inequality in Eq. (6.39) is violated, then $\hat{\mathbf{B}}$ and $\hat{\mathbf{C}}$ may drift to infinity with time. To accommodate these, one can set upper bounds $\mathcal{B}$ and $\mathcal{C}$ on $\|\hat{\mathbf{B}}\|$ and $\|\hat{\mathbf{C}}\|$, respectively. Thus the modified adaptation laws are:

$$\dot{\hat{\mathbf{B}}}^T = \begin{cases} -\Gamma_1^{-1}\mathbf{u}\mathbf{e}^T\mathbf{P}, & \text{if } \|\hat{\mathbf{B}}\| \leq \mathcal{B} \\ 0, & \text{otherwise} \end{cases} \tag{6.40}$$

$$\dot{\mathbf{C}} = \begin{cases} -\Gamma_2^{-1}\boldsymbol{\Phi}(\mathbf{x})\mathbf{e}^T\mathbf{P}, & \text{if } \|\hat{\mathbf{C}}\| \leq \mathcal{C} \\ 0, & \text{otherwise} \end{cases} \tag{6.41}$$

According to the above modified update laws, $\dot{V}$ is always negative semi-definite, and stability arguments are the same as in the case of $\epsilon = 0$. However, in the case where the bound in Eq. (6.39) is violated, the estimates of $\hat{\mathbf{B}}$, $\hat{\mathbf{C}}$ and $\mathbf{e}$ may increase as $\dot{V} > 0$, but all the quantities are still bounded due to the adaptation law in Eqs. (6.40) and (6.41).

It should be noted that Eq. (6.39) reiterates the importance of accurate approximation of nonlinear function $\mathbf{g}(.)$. According to the Stone-Weierstrass's approximation theorem, as $N \rightarrow \infty$, the approximation error $\epsilon \rightarrow 0$ over a compact Hausdroff space. However, in practice this is not possible, as these adaptation laws are based upon the assumption that all the parameters of the network can be optimized simultaneously. The global nature of the continuous map, $\mathbf{g}(.)$, can lead to globally-optimal network parameters which adequately minimize the approximation error, but not to the desired level. An alternative to global learning is local learning using local weight functions. The local learning algorithms involve estimation of network parameters using the observations in the local neighborhood of the operating point. Generally, the sizing of the local neighborhood is dictated by the support or domain of the weight functions. In Chapter 5, an approximation method is presented that enables a piecewise continuous approximation in an n-dimensional space using orthogonal polynomials and specially designed weight functions for overlapping the approximations in contiguous overlapping local regions to obtain the desired order of global continuity. Further, we have earlier shown that the introduction of local models and averaging of different local approximations improves the approximation accuracy for a continuous map. In the next section, those results will be extended for the dynamical system identification case so that the approximation error, ϵ, can be significantly reduced. The adaptive nature of this approximation approach can essentially guarantee a small ϵ, if low noise measurement density in space and time is available.

6.4.1.1 Adaption Law Derivation Using the GLO-MAP Model

As discussed in Chapter 5, the main idea of the GLO-MAP algorithm is a weighting function technique that generates a global family of overlapping preliminary approximations whose centroids of validity lie on the vertices of an n-dimensional grid, with vertices separated by a uniform step, h. These preliminary approximations are constructed so they represent the behavior in local hypercubes with a volume $(2h)^n$ centered on a typical vertex in the grid. A novel averaging process is developed in Ref. [62, 63] to determine a piecewise continuous global family of local least squares approximations, while having the freedom to vary the nature (e.g., degrees of freedom) of the local approximations. The continuity conditions are enforced by using a unique set of weighting functions in the averaging process. The weight functions are designed to guarantee the global continuity conditions while retaining near complete freedom on the selection of the generating local approximations.

To illustrate this approach let us first assume $n = 2$, and $g(x_1, x_2) : \mathcal{R}^2 \to \mathcal{R}^2$ is a continuous function which can be approximated by the GLO-MAP process according to the following equation:

$$\mathbf{g}(x_1, x_2) = w_{0,0}(x_1^{I_1}, x_2^{I_2})\mathbf{g}_{I_1,I_2}(x_1, x_2) + w_{0,1}(x_1^{I_1}, x_2^{I_2+1})\mathbf{g}_{I_1,I_2+1}(x_1, x_2) + \cdots$$
$$w_{1,0}(x_1^{I_1+1}, x_2^{I_2})\mathbf{g}_{I_1+1,I_2}(x_1, x_2) + \cdots$$
$$w_{1,1}(x_1^{I_1+1}, x_2^{I_2+1})\mathbf{g}_{I_1+1,I_2+1}(x_1, x_2)$$

$$= \underbrace{\left[\mathbf{g}_{I_1,I_2}(.) \cdots \mathbf{g}_{I_1+1,I_2+1}(.) \right]}_{\equiv\ \mathbf{F}_{2\times4}} \underbrace{\left\{ \begin{array}{c} w_{0,0}(x_1^{I_1}, x_2^{I_2}) \\ w_{0,1}(x_1^{I_1}, x_2^{I_2+1}) \\ w_{1,0}(x_1^{I_1+1}, x_2^{I_2}) \\ w_{1,1}(x_1^{I_1+1}, x_2^{I_2+1}) \end{array} \right\}}_{\equiv\ \mathbf{W}_{4\times1}} \qquad (6.42)$$

where $x_i^I = \frac{x_i - X_i^I}{h}$ is a local coordinate and X_i^I denotes grid point coordinates. Also, the weight functions are chosen such that these functions form a partition of unity so that they satisfy

$$\sum_{i_1=0}^{1} \sum_{i_2=0}^{1} w_{i_1 i_2}(\ ^{I_1+i_1}x_1, \ ^{I_2+i_2}x_N) = 1 \qquad (6.43)$$

Further, the local approximations, $\mathbf{f}_{I_1,I_2}(x_1, x_2)$, can be approximated by a set of orthogonal basis functions, $\boldsymbol{\Phi}$, as follows:

$$\mathbf{g}_{I_1,I_2}(x_1, x_2) = \mathbf{c}_{I_1,I_2}\boldsymbol{\phi}(x_1^{I_1}, x_2^{I_2}) \qquad (6.44)$$

Now, making use of Eq. (6.44), the matrix $\mathbf{F}$ in Eq. (6.42) can be rewritten as

$$\mathbf{F} = \underbrace{\left[\mathbf{c}_{I_1,I_2} \cdots \mathbf{c}_{I_1+1,I_2+1} \right]}_{\equiv\ \mathbf{C}_{2\times4N}}$$
$$\underbrace{\left[\begin{array}{cccc} \boldsymbol{\phi}(x_1^{I_1}, x_2^{I_2}) & O_{N\times1} & \cdots & O_{N\times1} \\ O_{N\times1} & \boldsymbol{\phi}(x_1^{I_1}, x_2^{I_2+1}) & O_{N\times1} & \vdots \\ \vdots & O_{N\times1} & \boldsymbol{\phi}(x_1^{I_1+1}, x_2^{I_2}) & O_{N\times1} \\ O_{N\times1} & O_{N\times1} & O_{N\times1} & \boldsymbol{\phi}(x_1^{I_1+1}, x_2^{I_2+1}) \end{array} \right]}_{\equiv\ \boldsymbol{\Phi}(.)_{4N\times4}} \qquad (6.45)$$

So, Eq. (6.42) reduces to

$$\mathbf{g}(x_1, x_2) = \mathbf{C}\boldsymbol{\Phi}(.)\mathbf{W} \qquad (6.46)$$

Now, using the approximation for $\mathbf{g}(.)$ given by Eq. (6.46), Eq. (6.23) reduces to:

$$\dot{\mathbf{x}}(t) = \mathbf{A}_l\mathbf{x}(t) + \mathbf{B}\mathbf{u}(t) + \mathbf{C}\underbrace{\mathbf{\Phi}(.)\mathbf{W}}_{\equiv \, \mathbf{\Psi}(.)} +\epsilon \tag{6.47}$$

Once again, the Fourier coefficient matrix, $\mathbf{C}$, and control effectiveness matrix, $\mathbf{B}$, are unknown and one can write

$$\dot{\hat{\mathbf{x}}}(t) = \mathbf{A}_l\hat{\mathbf{x}}(t) + \hat{\mathbf{B}}\mathbf{u}(t) + \hat{\mathbf{C}}\mathbf{\Psi}(.) \tag{6.48}$$

Let us define $\mathbf{e}(t) = \mathbf{x}(t) - \hat{\mathbf{x}}(t)$, and, time derivative of $\mathbf{e}(t)$ can be written as

$$\dot{\mathbf{e}}(t) = \mathbf{A}_l\mathbf{e} + (\mathbf{B} - \hat{\mathbf{B}})\mathbf{u}(t) + (\mathbf{C} - \hat{\mathbf{C}})\mathbf{\Psi}(.) + \epsilon \tag{6.49}$$

$$= \mathbf{A}_l\mathbf{e} + \tilde{\mathbf{B}}\mathbf{u}(t) + \tilde{\mathbf{C}}\mathbf{\Psi}(.) + \epsilon \tag{6.50}$$

Now to find adaptation laws for unknown parameters, let us consider the following Lyapunov function:

$$V = \frac{1}{2}\mathbf{e}^T\mathbf{P}\mathbf{e} + \frac{1}{2}Tr(\tilde{\mathbf{B}}\mathbf{\Gamma}_1\tilde{\mathbf{B}}^T) + \frac{1}{2}Tr(\tilde{\mathbf{C}}\mathbf{\Gamma}_2\tilde{\mathbf{C}}^T) \tag{6.51}$$

where $\mathbf{P}$ is a positive definite symmetric matrix. Now, taking the time derivative of V leads to the following equation:

$$\dot{V} = \frac{1}{2}\mathbf{e}^T\underbrace{\left(\mathbf{P}\mathbf{A}_l + \mathbf{A}_l^T\mathbf{P}\right)}_{\equiv \, -\mathbf{Q}}\mathbf{e} + \mathbf{e}^T\mathbf{P}\left(\tilde{\mathbf{B}}\mathbf{u}(t) + \tilde{\mathbf{C}}\mathbf{\Psi}(.) + \epsilon\right) + Tr\left(\tilde{\mathbf{B}}\mathbf{\Gamma}_1\dot{\tilde{\mathbf{B}}}^T\right)$$

$$+ Tr\left(\tilde{\mathbf{C}}\mathbf{\Gamma}_2\dot{\tilde{\mathbf{C}}}^T\right)$$

$$= -\frac{1}{2}\mathbf{e}^T\mathbf{Q}\mathbf{e} + Tr\left(\tilde{\mathbf{B}}\left[\mathbf{\Gamma}_1\dot{\tilde{\mathbf{B}}}^T + \mathbf{u}\mathbf{e}^T\mathbf{P}\right]\right) + Tr\left(\tilde{\mathbf{C}}\left[\mathbf{\Gamma}_2\dot{\tilde{\mathbf{C}}}^T + \mathbf{\Psi}(.)\mathbf{e}^T\mathbf{P}\right]\right)$$

$$+ \mathbf{e}^T\mathbf{P}\epsilon \tag{6.52}$$

Therefore, if the following adaptation laws are chosen for $\tilde{\mathbf{B}}$ and $\tilde{\mathbf{C}}$,

$$\dot{\tilde{\mathbf{B}}}^T = -\mathbf{\Gamma}_1^{-1}\mathbf{u}\mathbf{e}^T\mathbf{P} \tag{6.53}$$

$$\dot{\tilde{\mathbf{C}}}^T = -\mathbf{\Gamma}_2^{-1}\mathbf{\Psi}(\mathbf{x})\mathbf{e}^T\mathbf{P} \tag{6.54}$$

then $\dot{V}$ reduces to

$$\dot{V} = -\frac{1}{2}\mathbf{e}^T\mathbf{Q}\mathbf{e} + \epsilon^T\mathbf{P}\mathbf{e} \tag{6.55}$$

$$\Rightarrow \dot{V} \leq -\frac{1}{2}|\lambda_{min}(\mathbf{Q})|\|\mathbf{e}\|^2 + \|\epsilon\|\|\mathbf{P}\|\|\mathbf{e}\| \tag{6.56}$$

Therefore, $\dot{V}$ is negative definite if $\|\epsilon\| \leq \frac{|\lambda_{min}(\mathbf{Q})|\|\mathbf{e}\|}{2\|\mathbf{P}\|}$. The bounded stability of the tracking residual, $\mathbf{e}$, follows from the same arguments as outlined in the last section.

The adaptation laws presented in this chapter do not guarantee the convergence of the unknown control effectiveness matrix, $\mathbf{B}$, and Fourier coefficients matrix, $\mathbf{C}$, to their true values, but do ensure that the parameter estimation errors are bounded. The convergence of unknown parameters to their true value can only be guaranteed by satisfying the persistence of excitation conditions [79].

The generalization of Eq. (6.42) is

$$\mathbf{g}(X_1, \cdots, X_N) = \sum_{i_1=0}^{1} \sum_{i_2=0}^{1} \cdots \sum_{i_N=0}^{1} \left(w_{i_1, \cdots, i_N} \left({}^{I_1+i_1} x_1, \cdots, {}^{I_N+i_N} x_N \right) \right.$$
$$\left. \mathbf{g}_{I_1+i_1, \cdots, I_N+i_N} (X_1, \cdots, X_N) \right) \tag{6.57}$$

However, the expression for the adaptation laws for the Fourier coefficients and the control effectiveness matrices remain the same, except that now matrix $\mathbf{C}$ in Eq. (6.54) consists of the coefficients of 2^n neighboring approximations depending upon the value of $\mathbf{x}$.

Finally, we mention that the state vector, $\mathbf{x}$, is generally unknown, so the best available estimates of state vector, $\hat{\mathbf{x}}$, are used as state measurements. These estimates can be obtained by using the Kalman filter algorithm along with the best known linear system, as discussed in Section 6.3. The convergence of the nonlinear system identification algorithm, SysID 1, can be proved under a reasonable set of assumptions, which are captured in the following theorem:

Theorem 6. *If the linear system described by Eqs. (6.6) and (6.7) is fully observable, and tuning parameters $\mathbf{P}$ and $\mathbf{Q}$ are chosen in such a way that $\dot{V}$ described by Eq. (6.55) is negative definite, then $\|\hat{\mathbf{y}} - \mathbf{y}\| \rightarrow \|\mathbf{C}_l(\epsilon_1 + e_{lb})\|$. where ϵ_1 bounds the hidden state estimation error and $e_{lb} = 2\frac{\|\epsilon\|\|\mathbf{P}\|}{|\lambda_{min}(\mathbf{Q})|}$.*

Proof. The observability of the identified linear dynamic system guarantees that the hidden state, $\mathbf{x}$, can be detected from the given output, $\mathbf{y}$. In other words, the Kalman filter estimate, $\hat{\mathbf{x}}_b$, can be obtained in such a way that

$$\|\mathbf{x} - \hat{\mathbf{x}}_b\| \leq \epsilon_1 \tag{6.58}$$

Now, we just need to show that $\hat{\mathbf{x}}$ asymptotically converges to $\hat{\mathbf{x}}_b$. Eq. (6.55) tells us the important qualitative truth: If the system can be modeled with sufficiently small errors as we have modeled it in Eq. (6.47), then asymptotic convergence is assured. However, in the presence of modeling errors, the tracking errors will converge with errors bounded by the following value, as discussed in the previous section:

$$\|\mathbf{e}\| \rightarrow 2\frac{\|\epsilon\|\|\mathbf{P}\|}{|\lambda_{min}(\mathbf{Q})|} = e_{lb} \tag{6.59}$$

This means that state identification error $\mathbf{e}$ is always bounded by e_{lb}, which further implies that

$$\begin{aligned}
\|\mathbf{y} - \hat{\mathbf{y}}\| &= \|\mathbf{C}_l(\mathbf{x} - \hat{\mathbf{x}})\| \\
&= \|\mathbf{C}_l(\underbrace{\mathbf{x} - \hat{\mathbf{x}}_b}_{\epsilon_1} + \hat{\mathbf{x}}_b - \hat{\mathbf{x}})\| \\
&\rightarrow \|\mathbf{C}_l(\epsilon_1 + e_{lb})\|
\end{aligned}$$

$\square$

The above theorem gives us the lower bound for the system identification errors which can be reduced to a desired tolerance by the judicious selection of various tuning parameters.

6.4.2 Learning Algorithm for Output Model Perturbation Approach (SysID 2)

In the previous section, we have described the learning algorithm for the nonlinear system identification algorithm, SysID 1. The performance and computational cost of SysID 1 depends upon the dimensionality of the hidden state vector, $\mathbf{x}$. As the dimensionality of state vector, $\mathbf{x}$, increases, the number of terms required to approximate nonlinear function $\mathbf{g}(.)$ in Eq. (6.23) increases exponentially. For example, a total of 6 basis functions are required for second order approximation of $\mathbf{g}(.)$ in 2-D, while the number of required basis functions for second order approximation shoots to 66 for a 10-D state vector. In case of the GLO-MAP approximation, this number can rise even more depending upon the number of local approximations involved. This unexpected increase in the number of parameters makes SysID 1 computationally unattractive for the identification of structural mechanics systems where the number of participating modes runs to 10-20.

In Section 6.3, we described an alternate output model perturbation approach (SysID 2) to tackle the issue of dimensionality of the state vector $\mathbf{x}$. The overall architecture of the system identification algorithm, SysID 2, is depicted in Fig. 6.2. Once again, the basic idea of SysID 2 is to split the system identification process into linear and nonlinear identification processes. The linear system identification process is the same as in the case of SysID 1, and the main difference lies in the nonlinear identification process. In SysID 1, the linear dynamical model is perturbed by the nonlinear term $\mathbf{g}(.)$ to compensate for errors arising due to system nonlinearities; whereas in SysID 2, the algebraic nonlinear perturbation of best estimates of the linear approximation of the output vector is introduced to compensate for unknown system nonlinearity effects.

Consider the perturbed linear dynamical model, where $\mathbf{h}(\mathbf{x}_p)$ is a vector of nonlinear terms.

$$\dot{\mathbf{x}} = \mathbf{A}_l \mathbf{x} + \mathbf{B} \mathbf{u} \tag{6.60}$$

$$\mathbf{y} = \hat{\mathbf{y}}_l + \mathbf{h}(\mathbf{x}_p) \tag{6.61}$$

with

$$\hat{\mathbf{y}}_l = \mathbf{C}_l \hat{\mathbf{x}} + \mathbf{D}_l \mathbf{u} \tag{6.62}$$

We note that adopting a model for the nonlinear output perturbations $\mathbf{h}(\mathbf{x}_p)$ is the first step in this approach to SysID. Here, $\mathbf{A}_l \in \mathcal{R}^{n \times n}$ is a known Hurwitz matrix, and $\mathbf{B} \in \mathcal{R}^{n \times p}$ is a control effectiveness matrix. Once again, we mention that the matrix $\mathbf{A}_l$ is chosen in such a way that it captures the modal frequencies of interest and can be obtained by the ERA or any other suitable linear system identification algorithm as described in Section 6.3. The vector $\mathbf{x}_p \in \mathcal{R}^s$ is a dummy variable which can be chosen as a physical variable associated with the problem at hand, but is not necessarily the same as the hidden state vector, $\mathbf{x}_l$. For example, in the case of modeling of a flexible space structure, the system output vector consists of surface distortion measurements at various spatial points. Therefore, the dummy variable $\mathbf{x}_p$ can consist of the Cartesian coordinates (x, y, z) of a sensor-defined reference frame. In other words, the surface distortions can be modeled as a function of Cartesian coordinates.

The time history estimates of the estimated linear output vector, $\hat{\mathbf{y}}_l$, can be obtained by using the procedure described in Section 6.3, so the system identification problem can be redefined as:

System Identification Problem. *Given the time history estimates of the state vector, $\mathbf{x}(t)$, and control variable, $\mathbf{u}(t)$, find estimates of the unknown nonlinearity vector, $\mathbf{h}(.)$.*

Now, due to obvious reasons, as discussed in Chapter 5, we use the GLO-MAP algorithm to approximate $\mathbf{h}(.)$:

$$\mathbf{h}(X_1, \cdots, X_s) = \sum_{i_1=0}^{1} \sum_{i_2=0}^{1} \cdots \sum_{i_s=0}^{1} \left(w_{i_1, \cdots, i_s} \left({}^{I_1+i_1} x_1, \cdots, {}^{I_s+i_s} x_s \right) \right.$$

$$\left. \mathbf{h}_{I_1+i_1, \cdots, I_s+i_s}(X_1, \cdots, X_s) \right) \tag{6.63}$$

where $x_i^I = \frac{x_i - X_i^I}{h}$ is a local coordinate, and X_i^I denotes grid point coordinates. Also, the weight functions are chosen such that these functions are a partition of unity so that they satisfy

$$\sum_{i_1=0}^{1} \sum_{i_2=0}^{1} \cdots \sum_{i_s=0}^{1} w_{i_1, \cdots, i_s} \left({}^{I_1+i_1} x_1, \cdots, {}^{I_s+i_s} x_s \right) = 1 \tag{6.64}$$

Further, the local approximations, $\mathbf{h}_{I_1+i_1, \cdots, I_s+i_s}(.)$, can be represented by a set of locally-supported orthogonal basis functions, $\mathbf{\Phi}$, as discussed in Chapter 5.

Finally, we mention that the convergence of SysID 2 follows from the guaranteed convergence of the Kalman filter and the GLO-MAP algorithm to within bounds associated with the actual dynamics being captured by Eqs. (6.60), (6.61) and (6.63). Further, the system identification error can usually be reduced to a desired tolerance by the judicious selection of the number of local approximations and degree of basis functions for each local approximation.

Logically, one might anticipate that the state perturbation approach might be preferred. In reality, the model errors arose physically in the system dynamic model, and the output model perturbation approach might be preferred if the output measurement model is highly nonlinear, or some more general hybrid method. It is not possible to construct a theory to select the best approach; however, as has been evident in our simulations, the output model perturbation approach is significantly easier to implement.

6.5   Numerical Simulation

The proposed nonlinear system identification algorithms are tested on a variety of test cases mainly concerned with large space structures. This section presents some results from these studies.

6.5.1   Dynamic System Identification of Large Space Antenna

Space-Based Radar (SBR) systems envisioned for the future may be a constellation of spacecraft that provide persistent real-time radar images of the Earth environment through the identification and tracking of moving targets, high-resolution synthetic aperture radar imaging and collection of high-resolution terrain information. The accuracy of the information obtained from the SBR system depends upon many parameters, like the geometric shape of the antenna, permittivities of the media through which the radar wave is traveling, etc., our ability to compensate in real time implicitly depends on the accuracy of system identification. Therefore, the characteristics of the scattered wave received by the SBR antenna for a given frequency depend on the surface and geometric parameters of the radar. To apply necessary corrections for scattering of radar waves, precise knowledge of the SBR antenna is necessary. However, the transient excitation of the flexible dynamics mode necessitated by the need to slew the antenna makes the shape estimation problem more difficult. While a variety of surface models can be employed to represent the instantaneous shape, we consider the case that the surface is measured at discrete points and a dynamical model for shape estimation is desired. The objective of this section is to apply the system identification methodologies developed in this chapter to estimate the real time SBR antenna shape using only the discrete time measurements of the antenna surface.

For simulation purposes, the SBR antenna geometry is modeled in NAS-TRAN [73]. The antenna model consists of a total of seven panels as shown in Fig. 6.3. Each panel is assumed to be 100 m long in length and 200×250 m^2 in area. It is assumed that shape deflection measurements are available at 1,500 uniformly distributed spatial points at each time instant. The reference frame associated with (x, y, z) is embedded in a control rigid body, the inertial motion of that body is assumed to be measured directly by an inertial measurement unit. NASTRAN is used to generate mass, $\mathbf{M}$, and stiffness, $\mathbf{K}$, matrices for the antenna structure, and coordinate transformation matrix, $\mathbf{T}$, to transform the modal coordinates to physical coordinates, i.e., deflections along each axis.

$$\text{Modal Equations: } \mathbf{M}\ddot{\boldsymbol{\eta}} + \mathbf{K}\boldsymbol{\eta} = 0 \qquad (6.65)$$

$$\text{Transformation to Physical Coord.: } \mathbf{y} = \mathbf{T}\boldsymbol{\eta} \qquad (6.66)$$

where $\boldsymbol{\eta}$ and $\mathbf{y}$ represent modal and physical coordinates, respectively. The order of the FEM model was $1{,}500 \times 3 = 4{,}500$. If one tries to use a traditional ANN method to find a continuous map between system output and input space then the order of such a model will be equal to the order of the FEM model, i.e., 4,500 (which is not desirable in terms of computational efficiency!). However, order reduction methods can be used to reduce the dimension of the model state space to $10 - 30$. These equations, augmented with artificial damping and nonlinearities, are simulated using the MATLAB [74] environment to generate the measurement data for 50 seconds at 10 Hz frequency. For the purpose of this chapter, radial basis functions are used to simulate artificial nonlinearity with random magnitude and center.

To test the effectiveness of the two system identification algorithms, we consider two test cases. In the first test case, the measurement data are generated by exciting the first two modes, while in the second test case, the first five modes are excited to generate measurement data. Now, according to the procedure listed in Section 6.3, first the ERA algorithm is used to generate a linear dynamic model for the SBR antenna model. As expected, the ERA system gives us 4th and 10th order linear dynamic models for the first and second test cases, respectively. Finally, the nonlinear system identification algorithms are used to refine the linear model learned by the ERA algorithm.

Fig. 6.4(a) shows some of the simulated measurements for various points on the antenna surface corresponding to the first test case, and Fig. 6.4(b) shows the relative output error plots corresponding to the ERA identified model. We mention that relative output error, e_y, is defined as

$$e_y = \frac{\|\mathbf{y}_{True} - \mathbf{y}_{Est}\|}{\|\mathbf{y}_{True}\|} \qquad (6.67)$$

From these plots, it is clear that although ERA is able to capture the two

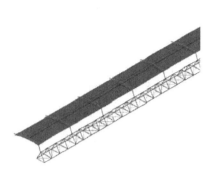 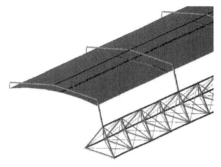

(a) NASTRAN SBR Antenna Model Consisting of Seven Panels   (b) Close-Up of One Panel

FIGURE 6.3
NASTRAN Model of the SBR Antenna.

modes of interest, there is significant error in estimating the true nonlinear output.

To model the effects of nonlinearities involved in the true dynamic model, the SysID 1 algorithm (*state model perturbation approach*) is employed according to the procedure listed in Section 6.4.1.1. Only one element is used to grid the estimated modal data, giving rise to a total of 16 local approximations. The orthogonal polynomial functions used to model the nonlinear function, $\mathbf{g}(.)$, are listed in Table 5.2.

$$\mathbf{g}_{I_1}(\mathbf{x}_l) = \mathbf{C}^T \mathbf{\Phi}(\mathbf{x}_l) \qquad (6.68)$$

where $\mathbf{x}_l \in \mathcal{R}^4$ consists of the ERA-identified modal coordinates. We mention that vector $\mathbf{\Phi}$ consists of only second and higher order terms in $\mathbf{x}_l$ to have the same linear dynamics as identified by the ERA algorithm. Therefore, $\mathbf{\Phi} \in \mathcal{R}^{10}$ and $\mathbf{C}$ is a 4×10 matrix of unknown Fourier coefficients. As a total of 16 local approximations are used to identify nonlinear function $\mathbf{g}(.)$, therefore, a total of $16 \times 10 = 160$ Fourier coefficients are required to be estimated. Initially, $\mathbf{C}$ is assumed to be a zero matrix and is adapted by using Eq. (6.54). Fig. 6.4(c) shows the adaptation plot of some of the Fourier coefficients.

In the case of the SysID 2 algorithm (*output model perturbation approach*), the unknown nonlinear function $\mathbf{h}(.)$ is approximated by the GLO-MAP algorithm, and dummy variable, $\mathbf{x}_p$, is assumed to consist of Cartesian coordinates (x, y, z). To approximate the SBR antenna shape at a particular time, the measurement data is modeled using a total of 64 finite element cells, 4 along each of the Cartesian coordinates, X, Y, and Z. Now, a continuous approximation of the SBR antenna shape for a particular cell is generated via a least squares procedure as listed in Chapter 5.

$$\hat{x}(x_l, y_l, z_l, t) = \sum_i \sum_j \sum_k a_{ijk} \phi_i(x_l) \phi_j(y_l) \phi_k(z_l), \; i + j + k \leq 2$$

$$\hat{y}(x_l, y_l, z_l, t) = \sum_l \sum_m \sum_n b_{lmn} \phi_l(x_l) \phi_m(y_l) \phi_n(z_l), \; l + m + n \leq 2$$

$$\hat{z}(x_l, y_l, z_l, t) = \sum_p \sum_q \sum_r c_{pqr} \phi_l(x_l) \phi_m(y_l) \phi_n(z_l), \; l + m + n \leq 2$$

Here, (x_l, y_l, z_l) denote the local Cartesian coordinates of a point predicted by using the linear system identified by the ERA algorithm. To learn the local approximations at each time, vision sensor measurements are processed sequentially. Initially, all Fourier coefficients are assumed to be zero and the corresponding covariance matrix initialized to 10^6 times the identity matrix.

Fig. 6.4(d) shows the relative output error plots for both of the system identification algorithms. From these plots, it is clear that the use of the nonlinear system identification algorithm reduces the estimation error by at least two orders of magnitude.

Further, Fig. 6.5(a) shows some of the true simulated measurements for various points on the antenna surface corresponding to the second test case, and Fig. 6.5(b) shows the relative output error plots corresponding to the ERA-identified model. From these plots, once again it is clear that although the ERA is able to approximately capture all five modes of interest, there is significant error in estimating the true nonlinear output.

Since the dimension of the linear state vector is 10, therefore, one needs a total of $2^{10} \times 55 = 56,320$ Fourier coefficients to learn the nonlinear function, $\mathbf{g}(.)$, using only one grid element. This high increase in the number of Fourier coefficients makes the use of the SysID 1 highly inefficient. Therefore, in the second test case, only the SysID 2 algorithm is used to refine the linear model learned by the ERA algorithm. Once again, the unknown nonlinear function, $\mathbf{h}(.)$, is approximated by the GLO-MAP algorithm, and dummy variable, $\mathbf{x}_p$, is assumed to consist of Cartesian coordinates (x, y, z). To approximate the SBR antenna shape at a particular time, the measurement data is modeled using a total of 64 finite element cells, 4 along each of the Cartesian coordinates, X, Y and Z.

Fig. 6.5(c) shows the relative output error plots for the SysID 2 system identification algorithms. Once again, the use of the nonlinear system identification algorithm reduces the estimation error by at least two orders of magnitude.

For the sake of simulations, it is implicitly assumed that actual surface deformation is sufficiently smooth and a relatively sparse set of measurements can provide support for the needed surface estimates. Of course, validating this assumption will be crucial in real applications and we did not attempt this. In the event that higher granularity is needed, of course, the present process can be readily refined by some combination of using more densely

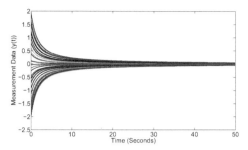

(a) Measured System Output Vector

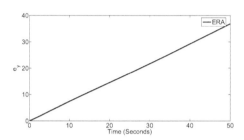

(b) Relative Error between the Measured and the ERA Estimated System Outputs

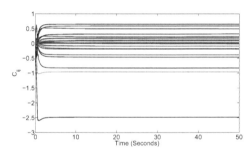

(c) Time History of Various Fourier Coefficients in the Case of SysID 1 Algorithm

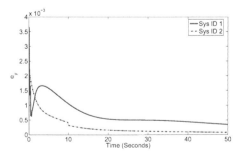

(d) Relative Error between the Measured and the Identified Nonlinear System Outputs

FIGURE 6.4

Non-Linear System Identification Results for Test Case 1.

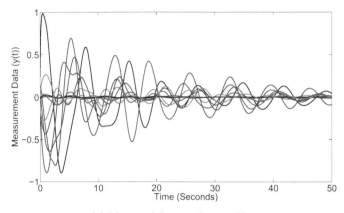

(a) Measured System Output Vector

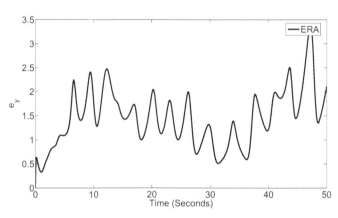

(b) Relative Error between the Measured and the ERA Estimated System Outputs

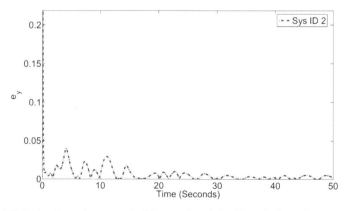

(c) Relative Error between the Measured and the Identified Nonlinear System Outputs

FIGURE 6.5

Non-Linear System Identification Results for Test Case 2.

spaced nodes or higher degree polynomial approximations. Finally, we mention that the simulation results presented in this section provide a basis for optimism regarding the utility of both of the algorithms. However, the effect of measurement space and time data frequency as well as sensor noise need to be considered in particular applications before making stronger conclusions about the utility of these algorithms.

6.6 Summary

A general methodology for nonlinear system identification is presented in this chapter. The method splits the nonlinear system identification process into two parts: (1) Linear system identification using ERA and (2) Nonlinear system identification using the GLO-MAP. We use the ERA-determined linear system state variable transformation to reduce nonlinear system state space dimensionality. We have found this approach to work well in the examples we have studied, but there is no theoretical guarantee that this approach to order reduction for a general nonlinear system will give the best order reduction. The GLO-MAP algorithm is used to learn the nonlinear correction term in the hidden state dynamical model and nonlinear transformation of measurement data for the *state model perturbation approach* (SysID 1) and the *output model perturbation approach* (SysID 2), respectively. A particularly attractive choice to model the nonlinear term is shown to be polynomial basis functions that are orthogonal with respect to the weight functions of the averaging process of the GLO-MAP algorithm. The adaption laws for different parameters of the GLO-MAP network are derived by using Lyapunov analysis in the case of SysID 1. The convergence of both of the algorithms is supported by a thorough analysis and demonstrated in the numerical study. The computational cost favors the output model perturbation approach (SysID 2) for high-dimensioned state space. The broad generality of the method, together with simulation results, provide a strong basis for optimism for the practical importance of these ideas. However, there remains an issue of uniqueness of the learned mathematical model; but again, we do not believe these open theoretical questions limit the usefulness of this approach for most engineering problems. We simply point out that proceeding with caution in the absence of theoretical justification is a necessary leap of faith until more theoretical progress can be made.

7

Distributed Parameter Systems

Each problem that I solved became a rule which served afterwards to solve other problems.

René Descartes

7.1 Introduction

The classical Finite Element Method (FEM) is a very pervasive and important approach to find the solution of Partial Differential Equations (PDEs). In the classical FEM, the approximate solution to PDEs is obtained by the discretization of the spatial domain into continuous volume/area elements. The local approximations for each element are obtained by the use of polynomial basis functions, which interpolate the solution and satisfy additional constraints, like exact interpolation at nodal points and inter-element continuity conditions. Generally, the polynomial degree, p, is fixed, and is dictated by the element type and the number of nodal points per element. The success of the classical FEM depends upon the approximation ability of the polynomial basis functions, and further improvement in the approximated solution can be achieved, for a given set of interpolation functions, only by refining the mesh size, h. In most cases, the degree of these basis functions is less than or equal to 2. For example, in the case of the triangular mesh with three nodes per element one can only use degree one polynomials for interpolation to satisfy necessary continuity requirements. Further, in some problems the use of non-polynomial basis functions may be desirable to achieve better accuracy. For example, in the case of the Helmholtz equation, the solution is known to be highly oscillatory in nature, and therefore, it may be desirable to use non-polynomial basis functions to approximate the exact solution. However, even though the analytical knowledge about local behavior of an exact solution is available, there are no convenient means to incorporate this knowledge in the conventional FEM solution. Also, the reliance of the conventional FEM on a mesh is not well suited to problems involving discontinuities and moving domains. Generally, to deal with moving domains and discontinuities in the conventional FEM methods, the original mesh is regenerated in each step of the evolution so that mesh lines are in accordance with the moving domain and discontinuities. However, this strategy of re-meshing at each stage can introduce numerous difficulties, such as the need to project the solution between meshes in successive stages of the problem, complexity in the computer program, and of course the computational burden associated with a large number of re-meshing steps.

The main objective of the meshless methods is to construct the PDEs solution entirely in terms of nodes in the absence of element connectivity. In the meshless FEM, nodal points can be added easily to the part of the domain where the solution is (expected to be) poor. In addition, since meshless methods use a non-element interpolation technique and a functional basis, therefore, the solution and its derivatives may be found directly where they are needed without additional interpolation errors. This makes the meshless FEM more flexible than the conventional FEM and a powerful tool to solve large classes of problems which are very awkward with a mesh-based conventional FEM.

Although the meshless methods are being emphasized in recent research literature [53, 90–95], the research efforts in this field have a long history [66, 90]. The main recent examples which follow the meshless FEM concepts are the Smoothed Particle Hydrodynamics Method (SPH), [90], the Element Free Galerkin Method (EFGM) [91],the Reproducing Kernel Particle (RKP) method [92], the Meshless Petrov-Galerkin (MLPG) method [53], the Partition of Unity Finite Element Method (PUFEM) [93] and the hp-cloud method [94]. An overview of various meshless methods can be found in Ref. [95]. Most of these methods, except the PUFEM, make use of the Moving Least Squares Approximation (MLSA) technique to find the expression for different local approximations. For the MLSA-based approach, the local shape functions are constructed through some local least squares data fitting, followed by the Galerkin discretization process to set up a linear system of equations. Finally, these systems of equations are solved for the solution value at specified nodal points. We mention that although successes have been many by using the MLSA-based meshless methods, there remain many drawbacks and the computational cost associated with these methods, depending upon the problem, may be high. An alternative to the MLSA is the *partition of unity* approach where the Galerkin discretization process is used directly to find the local shape function, instead of using some data fitting process followed by the Galerkin discretization. The main advantage of the PUFEM over the MLSA approach is that the PUFEM approach results in a continuous approximation of the exact solution, while the MLSA just provides the solution value at specified nodal points; an auxiliary interpolation process is required to obtain the solution value at any point other than the specified nodal points.

A common feature of all meshless methods is a weight function which is used to define the domain of integration for a particular node. The specially designed weight functions are positive with compact support, which dictates

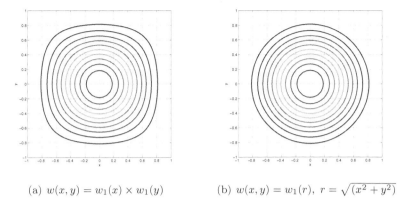

(a) $w(x,y) = w_1(x) \times w_1(y)$ (b) $w(x,y) = w_1(r)$, $r = \sqrt{(x^2 + y^2)}$

FIGURE 7.1

Different Shapes for Domain of Integration.

the domain of integration for a particular nodal point. We mention that the domain of integration defines the local region over which a local weak form associated with a particular node is valid, and it is analogous to the element space in the conventional FEM. The most commonly used sub-domains are circular, rectangular or elliptical in shape. Depending upon the input argument to weight functions, different shapes can be achieved for domain of integration. For example, in Fig. 7.1(a), a rectangular shape is obtained by the tensor product of 1-D weight functions, while in Fig. 7.1(b) the circular shape is obtained by selecting the input argument of the 1-D weight function to be the radial distance of the point from origin or nodal point in question. In the case of PUFEM, these weight functions need to satisfy an extra constraint: *partition of unity*. Note that the GLO-MAP weight functions are positive functions with compact support and form a partition of unity. As a consequence of this, they can be used in various meshless methods. The main advantage of using the GLO-MAP weight functions is that they are polynomial in nature (whereas most PUFEM weight function are not simple polynomials). Further, if one uses the polynomial functions orthogonal to the GLO-MAP weight function as the basis functions to locally approximate the solution, then many integrals can be evaluated accurately and easily without numerical integration.

In this chapter, attention is focused on the use of the GLO-MAP weight functions along with the Galerkin discretization process to establish a new algorithm for solving PDEs in an efficient manner. Modifications of the standard MLPG and PUFEM approaches are proposed using the GLO-MAP algorithm. We mention that the novel averaging process and orthogonal function approximation of the GLO-MAP algorithm differentiate it advantageously from its conventional counterparts.

The structure of this chapter is as follows: first, the MLSA-based Meshless Petrov-Galerkin (MLPG) method is described. Next, the use of the GLO-MAP algorithm is described in context with the PUFEM approach. Finally, numerical studies that compare the performance of various algorithms are summarized.

7.2   MLPG-Moving Least Squares Approach

In this section, the main characteristics of the Meshless Petrov-Galerkin algorithm are discussed and one should refer to Refs. [53, 96] for more detailed discussions on this method.

Let us consider the following linear PDE to be solved over global domain Ω with boundary Γ

$$\mathcal{L}u = f \tag{7.1}$$

and the following boundary conditions

$$u = \bar{u} \text{ on } \Gamma_u \tag{7.2}$$

$$\nabla u.\hat{\mathbf{n}} = \bar{q} \text{ on } \Gamma_q \tag{7.3}$$

where $\mathcal{L}$ is the general differential operator, u is the unknown function to be solved and f is the forcing term. Further, Γ_u and Γ_q are parts of the global boundary Γ where Dirichlet and Neumann boundary conditions are imposed, respectively. Finally, $\hat{\mathbf{n}}$ is the outward normal vector.

Like most FEM formulations, we approximate the unknown function u as $\hat{u}$, and write a generalized local weak form of Eq. (7.1) over a local sub-domain Ω_x:

$$\int_{\Omega_x} [\mathcal{L}\hat{u} - f]v_x d\Omega + \alpha \int_{\Gamma_{xu}} [\hat{u} - \bar{u}]v_x d\Gamma + \beta \int_{\Gamma_{xq}} [q - \bar{q}]v_x d\Gamma = 0 \tag{7.4}$$

where v_x is the test function associated with nodal point $\mathbf{x}$ that has a compact support $\Omega_x \subset \Omega$, also known as the domain of integration as shown in Fig. 7.2. The choice of the test function v_x determines the shape and size of the local domain Ω_x. Further, Γ_{xu} and Γ_{xq} are the boundary parts of the sub-domain Ω_x over which Dirichlet and Neumann boundary conditions are imposed, i.e., $\Gamma_{xu} = \Gamma_x \cap \Gamma_u$ and $\Gamma_{xq} = \Gamma_x \cap \Gamma_q$. α and β are penalty parameters used to approximately impose the Dirichlet and Neumann boundary conditions, respectively.

The approximation $\hat{u}$ of the unknown function, u, can be written as

$$\hat{u}(\mathbf{x}) = \sum_{i=1}^{N} \psi_i(\mathbf{x})\hat{u}_i \tag{7.5}$$

where N is the total number of nodal points used to discretize the domain Ω, $\hat{u}_i$ is the solution value at the ith nodal point, and $\psi_i(.)$ is the shape function associated with the ith nodal point with compact support Ω_i. The support Ω_i of the shape function ψ_i is known as the domain of definition in the literature [53], and is shown in Fig. 7.2. Finally, the substitution of Eq. (7.5) in Eq. (7.4) leads to the following set of linear equations in unknown variables $\hat{u}_i$:

$$\mathbf{K}\hat{\mathbf{u}} = \mathbf{f} \tag{7.6}$$

where $\mathbf{K}$ and $\mathbf{f}$ are given as

$$K_{ij} = \int_{\Omega_x} \mathcal{L}\psi_j(\mathbf{x})v_{x_j}d\Omega + \alpha \int_{\Gamma_{xu}} \psi_j(\mathbf{x})v_{x_i}d\Gamma + \beta \int_{\Gamma_{xq}} \nabla\psi_j(\mathbf{x})v_{x_i}d\Gamma \tag{7.7}$$

$$f_i = \int_{\Omega_x} fv_{x_i}d\Omega + \alpha \int_{\Gamma_{xu}} \bar{u}v_{x_i}d\Gamma + \beta \int_{\Gamma_{xq}} \bar{q}v_{x_i}d\Gamma \tag{7.8}$$

To obtain the expression for the shape function $\psi_i(.)$, the Moving Least Squares (MLS) fitting algorithm is adopted. This is also known as the local regression algorithm in the literature. The moving least squares approximation u^h of the unknown function u is written as

$$u^h(\mathbf{x}) = \phi^T(\mathbf{x})\mathbf{a}(\mathbf{x}) \tag{7.9}$$

where $\phi \in \mathcal{R}^m$ is a vector whose elements constitute a complete set of basis functions ϕ_i, and $\mathbf{a} \in \mathcal{R}^m$ is a vector of corresponding Fourier coefficients, a_i, which are a function of $\mathbf{x}$, instead of being constant as in the conventional Gaussian least squares approach. Furthermore, a set of nodal points $\{\mathbf{x}_i\}_{i=1,\cdots,M}$ are considered in the neighborhood Ω_M of $\mathbf{x}$, and the coefficient vector $\mathbf{a}(\mathbf{x})$ is obtained by minimizing the mean square error.

$$J = \sum_{i=1}^{M} w_i(\mathbf{x}, \mathbf{x}_i)\left(\phi^T(\mathbf{x}_i)\mathbf{a}(\mathbf{x}) - u_i^h\right)^2 \tag{7.10}$$

Here, u_i^h is the value of unknown function u at points $\mathbf{x}_i$ for which we want to solve. w_i is a weight function associated with the ith node such that $w_i(\mathbf{x}, \mathbf{x}_i) > 0$. In addition to the positivity, the weight function w_i also satisfies following properties:

1. The domain Ω_i of weight function w_i is a compact sub-space of Ω.

2. w_i is a monotonically decreasing function in $\|\mathbf{x} - \mathbf{x}_i\|$.

3. As $\|\mathbf{x} - \mathbf{x}_i\| \to 0$, $w_i \to \delta$.

According to the definition of weight function w_i, the domain of definition, Ω_m, of a point $\mathbf{x}$ is defined as collection of points for which $w_i(\mathbf{x}, \mathbf{x}_i) \geq 0$, $i =$

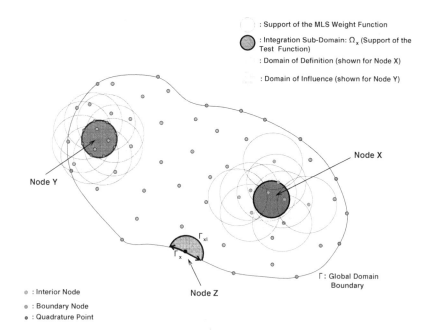

FIGURE 7.2
Illustration of the Domain of Definition, Domain of Influence and Domain of Integration for the MLPG Algorithm.

$1, 2, \cdots, M$. In other words, Ω_M can be defined as the union of sub-domains, Ω_i, i.e., $\Omega_M = \bigcup\limits_{i=1}^{M} \Omega_i$

Generally, a Gaussian weight function of the following form is used:

$$w_i(\mathbf{x}) = \begin{cases} \frac{e^{-(\|\mathbf{x}-\mathbf{x}_i\|/c)^2} - e^{-(r_i/c)^2}}{1 - e^{-(r_i/c)^2}}, & \|\mathbf{x}-\mathbf{x}_i\| \leq r_i \\ 0, & \|\mathbf{x}-\mathbf{x}_i\| > r_i \end{cases} \tag{7.11}$$

where r_i and c are parameters which dictate the size of domain of definition, Ω_m. As another possibility, a spline weight function is used for the MLS approximation in Ref. [53].

$$w_i(\mathbf{x}) = \begin{cases} 1 - 6\left(\frac{d_i}{r_i}\right)^2 + 8\left(\frac{d_i}{r_i}\right)^3 - 3\left(\frac{d_i}{r_i}\right)^4, & d_i = \|\mathbf{x}-\mathbf{x}_i\| \leq r_i \\ 0, & d_i > r_i \end{cases} \tag{7.12}$$

The first-order optimality condition for the loss function of Eq. (7.10) results in the following set of linear equations for $\mathbf{a}(\mathbf{x})$:

$$\mathbf{a}(\mathbf{x}) = \underbrace{(\boldsymbol{\Phi}^T \mathbf{W}(\mathbf{x})\boldsymbol{\Phi})^{-1}}_{\mathbf{A}(\mathbf{x})}\boldsymbol{\Phi}^T \mathbf{W}(\mathbf{x})\mathbf{u}^h \tag{7.13}$$

where $\mathbf{u}^h \in \mathcal{R}^M$ is a vector with entries u_i^h, and the matrices $\boldsymbol{\Phi}$ and $\mathbf{W}$ are given by

$$\boldsymbol{\Phi}_{ij} = \phi_j(\mathbf{x}_i) \tag{7.14}$$
$$\mathbf{W}_{ij} = w_i(\mathbf{x})\delta_{ij} \tag{7.15}$$

The necessary condition for the MLS solution to exist is that the rank of the matrix $\boldsymbol{\Phi}$ should be at least m. As a consequence of this, the domain of definition Ω_M should consist of at least m nodal points. Now, the MLS approximated solution, u^h, can also be expressed as

$$u^h(\mathbf{x}) = \sum_{i=1}^{n} \psi_i(\mathbf{x})u_i \tag{7.16}$$

where the shape function ψ_i is given by

$$\psi_i(\mathbf{x}) = \begin{cases} \sum_{i=1}^{m} \phi_j(\mathbf{x})(\mathbf{A}^{-1}(\mathbf{x})\boldsymbol{\Phi}\mathbf{W}(\mathbf{x}))_{ji} & w_i(\mathbf{x}) > 0 \\ 0 & w_i(\mathbf{x}) = 0 \end{cases} \tag{7.17}$$

Note that the shape function $\psi_i(\mathbf{x})$ vanishes at the nodal points where the weight function $w_i(\mathbf{x}) = 0$. The continuity of the shape function $\psi(.)$ depends upon the continuity of the weight function w_i and basis functions $\phi(.)$. Usually, the basis functions are chosen as m^{th} order polynomial functions in $\mathbf{x}$; however, one has the freedom to choose any set of basis functions depending upon the problem at hand. Note, for $m = 1$, the shape function $\psi(\mathbf{x})$ is given by the following expression and is known as Shepard's function:

$$\psi_i(\mathbf{x}) = \frac{w_i(\mathbf{x})}{\sum_{i=1}^{M} w_i(\mathbf{x})} \tag{7.18}$$

So, the continuity of the Shepard function depends solely upon the continuity of the weight functions. Note that the previously discussed GLO-MAP weight functions form a rigorous partition of unity, guarantee arbitrary order continuity, and satisfy all requirements of the MLS weight function. In particular, they do not require the normalization step of Eq. (7.18). The weight functions for the first four orders of continuity and their first derivatives, for 2-D approximations, are shown in Figs. 7.3, 7.4 and 7.5, respectively. The main advantage of using the GLO-MAP weight functions is that they are polynomial in nature, and further, if one uses the polynomial basis functions orthogonal to the GLO-MAP weight function to construct local approximations, then the shape function can be evaluated accurately and easily.

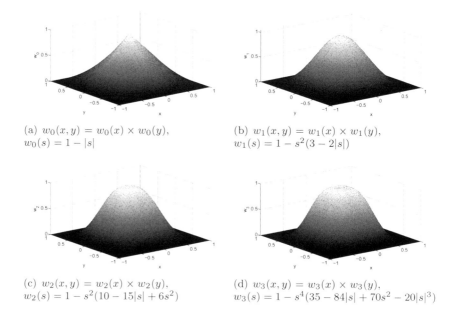

(a) $w_0(x,y) = w_0(x) \times w_0(y)$,
$w_0(s) = 1 - |s|$

(b) $w_1(x,y) = w_1(x) \times w_1(y)$,
$w_1(s) = 1 - s^2(3 - 2|s|)$

(c) $w_2(x,y) = w_2(x) \times w_2(y)$,
$w_2(s) = 1 - s^2(10 - 15|s| + 6s^2)$

(d) $w_3(x,y) = w_3(x) \times w_3(y)$,
$w_3(s) = 1 - s^4(35 - 84|s| + 70s^2 - 20|s|^3)$

FIGURE 7.3
Weight Functions for the First Four Orders of Continuity.

7.2.1   Poisson Equation

To illustrate the whole procedure of the MLPG approach, we consider the Poisson equation in 2-D space.

$$\nabla^2 u = f \text{ in } \Omega \tag{7.19}$$

$$u = \bar{u} \text{ on } \Gamma_u \tag{7.20}$$

$$u_{,n} = \bar{q} \text{ on } \Gamma_q \tag{7.21}$$

where $\nabla(.) = [\frac{\partial}{\partial x^2} + \frac{\partial}{\partial y^2}](.)$ is the Laplace operator, and n is the direction normal to the boundary of the domain. Analogous to Eq. (7.4), we write a generalized local weak form of the Poisson equation over a local sub-domain Ω_x as

$$\int_{\Omega_s} (\nabla^2 \hat{u} - f)v_x d\Omega - \alpha \int_{\Gamma_{xu}} (\hat{u} - \bar{u})v_x d\Gamma = 0 \tag{7.22}$$

where $\hat{u}$ is the trial function, v_x is the test function associated with nodal point x and Γ_{xu} is a part of the boundary Γ_u. Note that the second term in Eq. (7.22) is introduced to impose the essential boundary condition. According to Eq. (7.22), the trial function should be at least twice differentiable, i.e., $\hat{u} \in \mathcal{C}^2$, while the test function, v_x, should be a continuous function, i.e., $v_x \in \mathcal{C}^0$. However, using the fact that $(\nabla^2 \hat{u})v_x = (\hat{u}_{,i}v)_{x,i} - \hat{u}_{,i}v_{x,i}$ as well

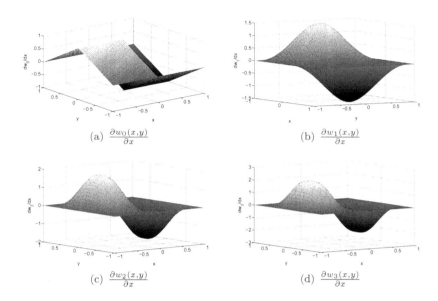

(a) $\frac{\partial w_0(x,y)}{\partial x}$

(b) $\frac{\partial w_1(x,y)}{\partial x}$

(c) $\frac{\partial w_2(x,y)}{\partial x}$

(d) $\frac{\partial w_3(x,y)}{\partial x}$

FIGURE 7.4
First Derivative of Weight Functions w.r.t. x for the First Four Orders of
Continuity.

as the divergence theorem, we can rewrite Eq. (7.26) such that both trial
function and test function are at least once differentiable, i.e., $v_x, \hat{u} \in \mathcal{C}^1$.

$$\int_{\partial\Omega_x} \hat{u}_{,i} n_i v_x d\Gamma - \int_{\Omega_x} (\hat{u}_{,i} v_{x,i} + f v_x) d\Omega - \alpha \int_{\Gamma_{xu}} (\hat{u} - \bar{u}) v_x d\Gamma = 0 \qquad (7.23)$$

Here, $\partial\Omega_x$ is the boundary of Ω_x, which can be divided into three parts:

$$\partial\Omega_x = \Gamma_{xu} + \Gamma_{xq} + \Gamma_{xI} \qquad (7.24)$$

where Γ_{xI} is the part of boundary $\partial\Omega_x$ which neither intersects Γ_u nor Γ_q.
Also, if we deliberately select a test function, v_x, such that it vanishes over the
boundary of sub-domain Ω_x, then the first term of Eq. (7.23) evaluated over
Γ_{xI} can be simplified. This can be accomplished easily by using the GLO-
MAP weight function as the test function. As mentioned earlier, the domain
of the test function v_x determines the domain Ω_x over which various integral
expressions of Eq. (7.23) should be evaluated. Here, we choose sub-domain
Ω_x to be a square centered at nodal point x. As test function v_x should be
at least $\mathcal{C}^1$, therefore, we choose this test function to be the second-order

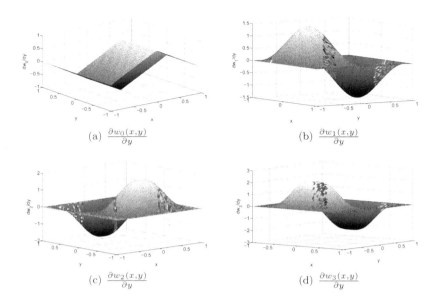

(a) $\frac{\partial w_0(x,y)}{\partial y}$

(b) $\frac{\partial w_1(x,y)}{\partial y}$

(c) $\frac{\partial w_2(x,y)}{\partial y}$

(d) $\frac{\partial w_3(x,y)}{\partial y}$

FIGURE 7.5

First Derivative of Weight Functions w.r.t. y for the First Four Order of Continuity.

GLO-MAP weight function.

$$v_x(x,y) = w_2(x) \times w_2(y) = \left[1 - \frac{|x|^3}{h}\left(10 - 15\frac{|x|}{h} + 6(\frac{x}{h})^2\right)\right]$$
$$\left[1 - \frac{|y|^3}{h}\left(10 - 15\frac{|y|}{h} + 6(\frac{y}{h})^2\right)\right] \quad (7.25)$$

Here, h is one half of the side of the square domain Ω_x. Now, using the fact that the test function vanishes over Γ_{xI}, Eq. (7.23) reduces to

$$\int_{\Omega_x} (\hat{u}_{,i}v_{x,i})d\Omega + \alpha \int_{\Gamma_{xu}} \hat{u}v_x d\Gamma - \int_{\Gamma_{xu}} q v_x d\Gamma = \int_{\Gamma_{xq}} \bar{q}v_x d\Gamma + \alpha \int_{\Gamma_{xu}} \bar{u}v_x d\Gamma$$
$$- \int_{\Omega_x} f v_x d\Omega \quad (7.26)$$

To obtain the algebraic equations from Eq. (7.26), the MLS approximation of Eq. (7.16) is used to approximate the trial function $\hat{u}$. To find the expression for the shape function, $\psi(.)$, we use the second-order GLO-MAP weight function, given by Eq. (7.25), and polynomial basis functions up to the second degree in both x and y. The polynomial functions for the MLS approximation are designed to be orthogonal to the MLS weight function and are shown in

Degree

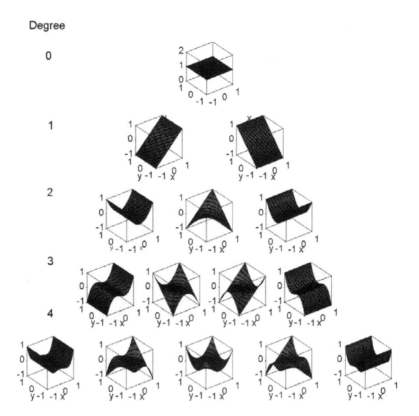

FIGURE 7.6
Two-Dimensional Polynomial Basis Functions Orthogonal to Weight Function
Given by Eq. (7.25).

Fig. 7.6. Finally, substitution of Eq. (7.16) into Eq. (7.26) for all nodes leads
to the following system of linear equations:

$$\mathbf{Ku} = \mathbf{f} \tag{7.27}$$

where the entries of the stiffness matrix, $\mathbf{K}$, and the load vector, $\mathbf{f}$, are given
as

$$\mathbf{K}_{ij} = \int_{\Omega_x} \psi_{j,k} v_{x,k}(\mathbf{x}, \mathbf{x}_i) d\Omega + \alpha \int_{\Gamma_{xu}} \psi_j v_x(\mathbf{x}, \mathbf{x}_i) d\Gamma - \int_{\Gamma_{xu}} \psi_{j,n} v_x(\mathbf{x}, \mathbf{x}_i) d\Gamma \tag{7.28}$$

$$\mathbf{f}_i = \int_{\Gamma_{xq}} \bar{q} v_x(\mathbf{x}, \mathbf{x}_i) d\Gamma + \alpha \int_{\Gamma_{xu}} \bar{u} v_x(\mathbf{x}, \mathbf{x}_i) d\Gamma - \int_{\Omega_x} f v_x(\mathbf{x}, \mathbf{x}_i) d\Omega \tag{7.29}$$

Note, theoretically, as long as the union of all local domains covers the global domain, i.e., $\cup\Omega_s \supset \Omega$, the equilibrium equation and the boundary conditions will be satisfied in the global domain and on its boundary, respectively. To ensure this, we choose parameter h to be the minimum distance of nodal point $\mathbf{x}$ from all other nodal points. Also, since we are using second-order polynomials for the MLS approximation, we need to make sure that there are at least 6 nodal points in the domain of definition, Ω_M, associated with each nodal point. To ensure this, we choose the support of Ω_i to be $6h$. The implementation of the MLPG method can be carried out according to the following steps and is illustrated in Fig. 7.7:

1. Choose a finite number of nodes to discretize the global domain, Ω, and global boundary, Γ.

2. Determine the local sub-domain, Ω_x, and its corresponding local boundary, $\partial\Omega_x$, for each node.

3. Loop over all nodes located inside the global domain and at the global boundary, Γ.

    (a) Determine Gaussian quadrature points, $\mathbf{x}_Q$, in the domain of integration, Ω_x, and its boundary, $\partial\Omega_x$.

    (b) Loop over the quadrature points, $\mathbf{x}_Q$, in the sub-domain, Ω_x and on the local boundary, $\partial\Omega_x$.

        i. Determine nodal points $\mathbf{x}_i$ such that $w_i(\mathbf{x}_Q, \mathbf{x}_i) > 0$.

        ii. Using the MLS approximation for the trial function evaluate numerical integrals in Eqs. (7.28) and (7.29).

        iii. Assemble contributions to the linear system for all nodes in $\mathbf{K}$ and $\mathbf{f}$.

    (c) End loop over quadrature points.

4. End node loop.

5. Solve the linear system for the fictitious nodal values $\hat{u}_i$.

To show the effectiveness of the method discussed in this section, we assume $f = -1$ in Eq. (7.19) and the following boundary conditions on a square domain of unit length:

$$u(x, 1) = u(1, y) = 0 \tag{7.30}$$

$$\frac{\partial u}{\partial x}\Big|_{(0,y)} = \frac{\partial u}{\partial y}\Big|_{(x,1)} = 0 \tag{7.31}$$

The exact analytical solution to this boundary value problem is given by the following equation [97]:

$$u(x, y) = \frac{1}{2}\left[(1 - y^2) + 4\sum_{n=1}^{\infty}\frac{(-1)^n \cos\alpha_n y \cosh\alpha_n x}{\alpha_n^3 \cosh\alpha_n}\right] \tag{7.32}$$

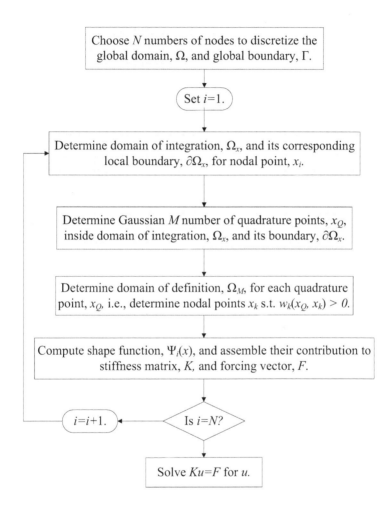

FIGURE 7.7
Flowchart for the MLPG Algorithm.

with

$$\alpha_n = \frac{1}{2}(2n - 1)\pi \qquad (7.33)$$

Fig. 7.8 shows the plots of the true solution surface and various partial derivatives of the true solution. The different integrals appearing in Eqs. (7.28) and (7.29) are evaluated numerically using the Gauss quadrature method. Regular meshes of different sizes are considered to study the convergence and accuracy of the method. In all of the cases, the computed solution is tested on a total of 2,500 uniformly distributed points inside a unit square. Fig. 7.9 shows the plot of relative error, e, with respect to mesh size, h, for linear, quadratic and cubic basis functions.

$$e = \frac{\sqrt{(u - \hat{u})^2 + (\frac{\partial u}{\partial x} - \frac{\partial \hat{u}}{\partial x})^2 + (\frac{\partial u}{\partial y} - \frac{\partial \hat{u}}{\partial y})^2}}{\sqrt{u^2 + \frac{\partial u}{\partial x}^2 + \frac{\partial u}{\partial y}^2}} \qquad (7.34)$$

As expected, the relative error decreases with the decrease in mesh size and the increase in the order of basis functions. Also, it is clear that the MLPG converges as might be expected to reasonably accurate results for the solution and its derivatives. Note that the improvement in accuracy between linear and cubic shape functions is approximately one order of magnitude for a fixed h.

7.2.2   Comments on the MLPG Algorithm

Although the MLS shape function used in the MLPG algorithm reproduces the true solution at nodal points in accordance with the least squares principle, there are several significant drawbacks associated with this approach.

1. The first disadvantage of the MLPG approach is that for each point under consideration a new linear system must be solved to find the value of the shape functions, ψ_i, and hence the value of the approximated solution, $\hat{u}$. This is a computationally burdensome task.

2. As mentioned earlier, the smoothness of the shape functions, ψ_i, and hence the smoothness of the approximated solution, $\hat{u}$, is directly related to the smoothness of both the basis functions and the weight functions used in the MLS approximation. As a consequence of this, the approximated solution, $\hat{u}$, is continuous up to an arbitrary order p over the whole domain Ω, if shape functions corresponding to all the nodes are continuous up to same order p. This is possible if the same set of basis functions is used in each local domain, Ω_M. In other words, the basis functions of the ith local region cannot be chosen independently from the basis function of the jth local region. So, one cannot increase the degree of approximation in a particular local region arbitrarily to reduce the approximation errors to a desired tolerance.

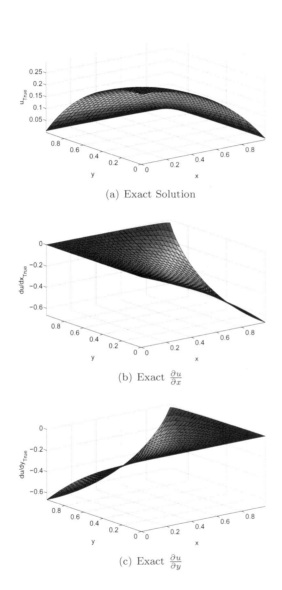

(a) Exact Solution

(b) Exact $\frac{\partial u}{\partial x}$

(c) Exact $\frac{\partial u}{\partial y}$

FIGURE 7.8
Exact Solution to Poisson's Equation.

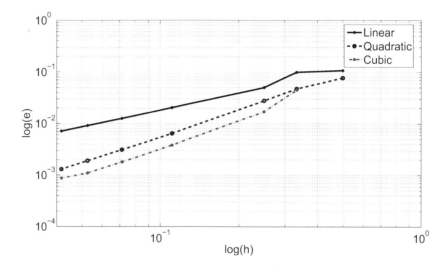

FIGURE 7.9
Relative Error for Poisson's Equation Using the MLPG Method.

3. The shape functions ψ_i fails to have the selective property known as *partition of unity*, i.e., $\sum_i \psi_i = 1$. Hence, u_i does not have the interpretation of the nodal value of $\hat{u}$.

$$u^h(\mathbf{x}_i) \neq u_i \qquad (7.35)$$

4. Even though the basis functions used to approximate u can be polynomial in nature, shape function ϕ_i will not generally be polynomial in nature, which makes numerical integration more difficult than if all functions in the integrands were polynomials.

7.3   Partition of Unity Finite Element Method

In the previous sections, we discussed the MLPG algorithm to solve partial differential equations. In the MLPG methods, the local shape functions are constructed with the help of methods from data fitting, and further, these shape functions are used in a Galerkin discretization process to set up a linear system of equations. Finally, these systems of equations need to be solved efficiently. Beside solving the final system of linear equations, one also needs to solve a new system of linear algebraic equations to evaluate the shape function value at any point which increases the computational cost and also restricts the accuracy that can be achieved. In this section, we directly use the Galerkin discretization process to find the local shape function, instead of using some data fitting process followed by the Galerkin discretization process.

The partition of unity viewpoint for the meshless FEM has been developed by Babuška and Melenk [93], whereas the partition of unity method for generalized piecewise continuous approximation was developed earlier by Junkins et al. [66]. The partition of unity is a mathematical paradigm in which a domain Ω is covered by overlapping sub-domains Ω_s, each of which is associated with a function f_s, which is non-zero over Ω_s, and has the property that

$$\sum_s f_s(\mathbf{x}) = 1 \qquad (7.36)$$

The summation is taken over all functions overlapping Ω_s. We mention that the partition of unity condition, given by Eq. (7.36), is identical to the zeroth order consistency condition for the functions $f_s(\mathbf{x})$, i.e., function $f_s(.)$ can reproduce the constant function exactly. This is accomplished by introducing local approximations within each Ω_s. For example, if the function values and the derivatives are given at the nodal points $\mathbf{x}_i$ then one can choose the function

$$\hat{u}(\mathbf{x}) = \sum_i f_i(\mathbf{x}) T_i(\mathbf{x}) \qquad (7.37)$$

where $T_i(.)$ are the Taylor polynomials of function $u(\mathbf{x})$ centered about the nodal point $\mathbf{x}_i$, and thus involve the value of the function and its derivatives evaluated at the nodal point, $\mathbf{x}_i$. Note the further consistency condition: If all of the approximations $T_i(\mathbf{x})$ were exact (i.e., $T_i(\mathbf{x}) = u(\mathbf{x})$), then the estimate in Eq. 7.37 also yields true for $u(\mathbf{x})$. We can use this observation to prove Eq. (7.37) is an unbiased estimation process, if T_i are unbiased. According to Eq. (7.37), there are two ways to improve the performance of the approximation $\hat{u}(.)$: One is to improve the order of the consistency of the shape functions $f_i(.)$; the other is to directly improve the consistency orders of the polynomial $T_i(.)$. The first one is difficult to achieve due to the partition of unity constraint given by Eq. (7.36). Alternatively, Shepard's function, $(f(\mathbf{x}) = \frac{f_i(.)}{\sum_i f_i(.)})$, is used as the zeroth order shape function, and the polynomial functions for the local approximation are chosen as

$$T_i(\mathbf{x}) = \sum_{j=1}^{m} \phi_{ij}(\mathbf{x}) a_{ij} \qquad (7.38)$$

where a_{ij} are the Fourier coefficients corresponding to various polynomial basis functions denoted by $\phi_{ij}(.)$. Now, the approximated function for the method of partition of unity can be written as

$$\hat{u}(\mathbf{x}) = \sum_{i=1}^{N} \sum_{j=1}^{m} f_i(\mathbf{x}) \phi_{ij}(\mathbf{x}) a_{ij} \qquad (7.39)$$

Using the above-described *partition of unity* paradigm, with compactly supported sub-domains forming a cover for the solution domain, a finite element approximation $\hat{u}$ of the function u can be constructed by setting up a conformal space of shape functions in Ω. This is accomplished by introducing basis functions, ϕ_{ij}, within each Ω_s. The basis functions, ϕ_{ij}, may be chosen from the space of polynomials or they may be *special functions*, based on prior knowledge about the problem. For example, if the solution of the PDE in question is known to be highly oscillatory, appropriate harmonic functions with spatial frequencies chosen based on knowledge of the system can be used in the basis set. Such special functions may be introduced either by themselves or to supplement a previously existing polynomial basis. This aspect, called "basis enrichment," is one of the greatest advantages of the PUFEM because it allows the use of local functions of different form and number in the individual sub-domains. While such freedom provides great flexibility and can immensely improve the approximability, it generally prevents the basis functions from constituting a conformal space, i.e., the inter-element continuity of the approximation is not ensured. This task is accomplished by the PU pasting functions, f_i, which merge together the various local approximations. Figure 7.10 illustrates the process of shape function construction in the PUFEM algorithm. In these figures, the basis functions (ϕ_{ij}) have been drawn using bold lines, and the PU pasting functions (f_i) using light lines. Also, all functions corresponding to odd-numbered nodes are drawn with solid lines and those corresponding to even-numbered nodes with dashed lines. The 1-D domain $[-1, 1]$ is discretized using 5 sub-domains with *tent*-functions in Fig. 7.10(a), and C^1 functions in Fig. 7.10(b). The use of quadratic polynomials as basis functions has been shown in all of the sub-domains. In addition, a sinusoidal function (which *enriches* the existing polynomial basis) has been introduced locally only in the third sub-domain (corresponding to the highlighted Node #3). Clearly, these basis functions do not form a conformal space on their own. However, when these functions are multiplied with the PU functions of the corresponding nodes, the resulting functions satisfy inter-element continuity and we refer to the product as *pasted basis functions*, or shape functions (see Figs. 7.10(c) and 7.10(d)).

$$\Psi_{sj} = f_s \phi_{sj}, \qquad j = 1, \ldots, Q_s \qquad (7.40)$$

In essence, the above approach delegates the burden of enforcing inter-element continuity to the PU functions so that the user is free to select basis functions purely on the criteria of local approximability. In comparison, in the conventional FEM the basis functions are the same as the shape functions; therefore they need to form a conformal space on their own, which limits their range of selection. Now, if Shepard's function is used as the shape function as suggested in Ref. [93], then the consistency in the above equation depends on the order of polynomial basis functions, $\phi(.)$. Further, note that Shepard's function provides the partition of unity and hence the compact support for the local approximation. The coefficients a_{ij} are the unknowns and can be found directly by using the Galerkin discretization process instead of using a data fitting algorithm. Therefore, one needs at least m test functions per node to obtain a sufficient number of equations to determine the unknown Fourier coefficients.

To illustrate the whole procedure, let us consider a general linear PDE given by Eq. (7.1), for which a generalized local weak form over a local sub-domain $\Omega_s \subset \Omega$ can be written as

$$\int_{\Omega_s} [\mathcal{L}\hat{u} - f]v_s d\Omega + \alpha \int_{\Gamma_{su}} [\hat{u} - \bar{u}]v_s d\Gamma + \beta \int_{\Gamma_{sq}} [q - \bar{q}]v_s d\Gamma = 0 \qquad (7.41)$$

where v_s represents the test function for sub-domain Ω_s and $\hat{u}$ is the approximated solution for Eq. (7.1), which is valid over sub-domain Ω_s. Further, substituting for $\hat{u}$ from Eq. (7.39) in Eq. (7.41), we get

$$\int_{\Omega_s} [\mathcal{L}\sum_{s=1}^{N}\sum_{j=1}^{m} f_s(\mathbf{x})\phi_{sj}(\mathbf{x})a_{sj} - f]v_s d\Omega + \alpha \int_{\Gamma_{su}} [\sum_{s=1}^{N}\sum_{j=1}^{m} f_s(\mathbf{x})\phi_{sj}(\mathbf{x})a_{sj} - \bar{u}]v_s d\Gamma$$

$$+ \beta \int_{\Gamma_{sq}} [q - \bar{q}]v_s d\Gamma = 0 \qquad (7.42)$$

To find unknown Fourier coefficients a_{sj}, we need at least m test functions per nodal point. Now, choosing test functions to be same as the trial functions, $\{f_s\phi_{si}\}_{i=1}^{m}$, we get the following set of algebraic equations for Fourier coefficients a_{si}:

$$\mathbf{Ka} = \mathbf{F} \qquad (7.43)$$

where $\mathbf{a}$ is a vector consisting of mN Fourier coefficients. Further, stiffness matrix, $\mathbf{K}$, and forcing vector, $\mathbf{F}$, are given as

$$K_{ij} = \int_{\Omega_s} \mathcal{L}f_i(\mathbf{x})\Phi_i(\mathbf{x})f_j\Phi_j(\mathbf{x})d\Omega + \alpha \int_{\Gamma_{su}} f_i(\mathbf{x})\Phi_i(\mathbf{x})f_j\Phi_j(\mathbf{x})d\Gamma$$

$$+ \beta \int_{\Gamma_{sq}} \frac{\partial f_i(\mathbf{x})\Phi_i(\mathbf{x})}{\partial n}f_j(\mathbf{x})\Phi_j d\Gamma \qquad (7.44)$$

$$F_j = \int_{\Omega_s} f f_j\Phi_j(\mathbf{x})d\Omega + \alpha \int_{\Gamma_{su}} \bar{u}f_j\Phi_j(\mathbf{x})d\Gamma + \beta \int_{\Gamma_{sq}} \bar{q}f_j\Phi_j(\mathbf{x})d\Gamma \qquad (7.45)$$

where $\Phi = \{\phi_s\}_{1=i}N$ is a vector consisting of basis functions for each sub-domain Ω_s. It should be noted that in this case the dimension of the stiffness matrix, $\mathbf{K}$, is $Nm \times Nm$, which is m times the dimension of the stiffness matrix in the case of the MLPG algorithm. However, in the case of the MLPG algorithm one needs to solve a new system of linear equations to find m Fourier coefficients for each local approximation. Also, in the case of the

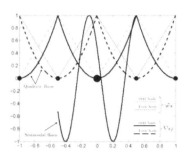

(a) PUFEM Basis Functions and the C^0 PU Function

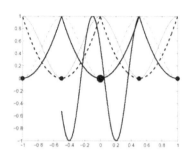

(b) PUFEM Basis Functions and the C^1 PU Function

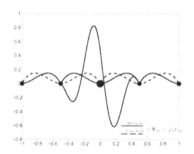

(c) PUFEM Shape Functions with C^0 Pasting

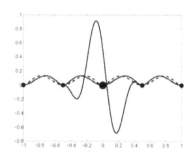

(d) PUFEM Shape Functions with C^1 Pasting

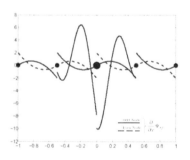

(e) Derivative of the Shape Functions with C^0 Pasting

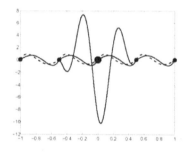

(f) Derivative of the Shape Functions with C^1 Pasting

FIGURE 7.10
Shape Function Construction in PUFEM.

MLPG algorithm, the computation of the solution and its derivative at a point other than the nodal point also involves the solution of a system of linear algebraic equations, whereas in the case of the PUFEM approach there is no extra computational burden associated with these calculations once one has solved Eq. (7.43) for Fourier coefficients.

It should be noted that the PU functions are of key importance in the PUFEM algorithm. They bring about the implicit domain discretization, merge together the various local approximations by performing an unbiased average, and determine their order of continuity across the local boundaries [93]. The order of continuity of the shape functions across the local sub-domains is inherited from the continuity of the PU functions [93]. Thus, the *tent*-functions, which are C^0 continuous, lead to shape functions whose derivatives are not continuous (Fig. 7.10(e)). On the other hand, the use of the PU functions with C^1 continuity leads to C^0 derivatives (Fig. 7.10(f)), which is an advantage. Because of the requirement of the PU constraint, it is generally a difficult task to construct PU functions that enforce continuity of any desired order. Shepard's function has been used to enforce the PU constraint. However, these functions are generally very difficult to integrate. Apparently the use of higher-order polynomials as PU functions, which could be automatically generated depending on the specified order of continuity of the approximation, has not heretofore been explored in the PUFEM literature. However, the GLO-MAP weight functions *are of polynomial form*, satisfy the PU constraint, and have compact support, thus satisfying the requirements for a PU, which makes them extremely attractive weight functions for the PUFEM approach. The idea behind the GLO-MAP weights is amazingly simple. Given a node belonging to a discretized domain, the polynomial function of the lowest degree, which assumes the value unity at the concerned node and decays to zero at all its neighboring nodes with the specified degree of smoothness, satisfies the property of partition of unity on the global domain Ω. In the PUFEM framework, these weights come as an invaluable construct because of their several relevant properties:

1. Polynomial form: By virtue of their polynomial form, GLO-MAP weights are very easy to integrate. Additionally, if polynomial bases are used, *the resulting weak form integrals can be evaluated analytically.*

2. They satisfy the PU property. It is very easy to prove the fulfillment of this constraint when the GLO-MAP weights are written in local co-ordinates centered at the corresponding nodes and scaled with the inter-nodal distance along each dimension, $h^{(i)}$. This implies that in the local coordinates, the central node is at the centroid of an N-hypercube and all its neighboring nodes are at the various 2^N vertices. The value of the GLO-MAP weights are 1 and 0, respectively, at these locations.

3. They can provide any desired order of continuity across sub-domain boundaries. This is very useful in applications which require the solution

derivatives to satisfy certain error bounds.

4. Easy extension to higher dimensions: It is surprisingly easy to construct GLO-MAP weights in higher dimensions. A simple continued product (tensor product) of 1-D weights written along the various dimensions gives the weight function in the higher dimensional space which satisfies all the properties mentioned above, e.g., $w_{(2)}(x, y) = w_{(2)}(x)w_{(2)}(y)$, i.e., a GLO-MAP weight in 2-D providing C^2 continuity is simply the continued product of two 1-D weights providing the same level of smoothness.

In summary, the generality provided by GLO-MAP weight functions and their easy extension to N-dimensions opens up the path for the implementation of the PUFEM algorithm to solve high-dimensional PDEs [98]. Furthermore, if basis functions orthogonal to these weight functions are used, we obtain an improvement in the condition number of the stiffness matrix, $\mathbf{K}$, and many of the analytical integrals, e.g., Eqs. (7.62)−(7.66), become trivial to compute [65]. A limitation of these functions is that in order to satisfy the PU constraint, the nodes must be aligned as if on a rectangular grid. However, it is possible to have variable inter-nodal separation between the various adjacent nodes along any particular dimension. This restriction implies that the PUFEM algorithm can be applied directly only to PDEs defined on N-hypercuboids. Domains of all other shapes would require a transformation into a hypercuboid.

7.3.1 Poisson Equation

To illustrate the PUFEM ideas, we again consider the Poisson equation in 2-D space:

$$\nabla^2 u = f \text{ in } \Omega \tag{7.46}$$

$$u = \bar{u} \text{ on } \Gamma_u \tag{7.47}$$

$$u_{,n} = \bar{q} \text{ on } \Gamma_{uq} \tag{7.48}$$

where $\nabla(.)$ is the Laplace operator, and n is the direction normal to the boundary of the domain. As discussed earlier, after some algebraic manipulations and making use of the divergence theorem, we get the following weak form equation from Eq. (7.46):

$$\int_{\partial\Omega_s} \hat{u}_{,i} n_i v_s d\Gamma - \int_{\Omega_s} (\hat{u}_{,i} v_{x,i} + f v_s) d\Omega - \alpha \int_{\Gamma_{su}} (\hat{u} - \bar{u}) v_s d\Gamma = 0 \tag{7.49}$$

where $\partial\Omega_s$ is the boundary of Ω_s, $\hat{u}$ is the trial function, approximated by the GLO-MAP process and v_s is the test function with support equal to the local sub-domain Ω_s. Further, we deliberately divide the boundary term $\partial\Omega_s$ in three parts, Γ_{su}, Γ_{sq}, and Γ_{sI}, where Γ_{su} and Γ_{sq} are the parts of the global boundary on which Dirichlet and Neumann boundary conditions are defined, and Γ_{sI} is the part of $\partial\Omega_s$ which does not intersect global boundary Γ. Now, if we deliberately select a test function, v_s, such that it vanishes over Γ_{sI}, then the first term of Eq. (7.49) can be simplified as discussed previously in this chapter. This can be easily accomplished by using the GLO-MAP weight function as the test function. Further, the substitution of Eq. (7.39) in Eq. (7.49) leads to the following system of linear equations:

$$\mathbf{Ka} = \mathbf{f} \qquad (7.50)$$

where

$$\mathbf{K}_{ij} = \int_{\Omega_s} ((f_j\phi^T)_{,k}v_{i,k})d\Omega + \alpha \int_{\Gamma_{su}} (f_j\phi^T - (f_j\phi^T)_{,k}n_{,k})v_i d\Gamma$$

$$- \int_{L_s} (f_j\phi^T)_{,k}v_i d\Gamma \qquad (7.51)$$

$$\mathbf{f}_i = \int_{\Gamma_{sq}} \bar{q}v_i d\Gamma + \alpha \int_{\Gamma_{su}} \bar{u}v_i d\Gamma - \int_{\Omega_s} fv_i d\Omega \qquad (7.52)$$

where $v_i = f_i\phi$ is the test function associated with the ith node. Note that if one uses the GLO-MAP weight functions as the partition of unity functions, f_i, and the corresponding set of orthogonal polynomials as basis functions, ϕ_i, then all integral terms in Eqs. (7.51) and (7.52) are polynomial in nature which can be evaluated analytically without further approximation. The implementation of the PUFEM can be carried out according to the following routine:

1. Choose a finite number of nodes to discretize the global domain, Ω, and global boundary, Γ.

2. Determine the local sub-domain, Ω_x, and its corresponding local boundary, $\partial\Omega_x$, for each node.

3. Assign partition of the unity weight function, $w_i(\mathbf{x}, \mathbf{x}_i)$, and basis functions, ϕ_i, with each node point, $\mathbf{x}_i$.

4. Loop over all nodes located inside the global domain and at the global boundary, Γ.

   (a) Determine Gaussian quadrature points, $\mathbf{x}_Q$, in the domain of integration, Ω_x, and its boundary, $\partial\Omega_x$.

   (b) Loop over quadrature points, $\mathbf{x}_Q$, in the sub-domain, Ω_x and on the local boundary, $\partial\Omega_x$.

      i. Determine nodal points, $\mathbf{x}_k$, such that $w_i(\mathbf{x}_i, \mathbf{x}_k) > 0$.

      ii. Evaluate numerical integrals in Eqs. (7.51) and (7.52).

        iii. Assemble contributions to the linear system for all nodes in $\mathbf{K}$ and $\mathbf{f}$.

   (c) End loop over quadrature points.

5. End node loop.

6. Solve the linear system for Fourier coefficients $\mathbf{a}$.

To show the effectiveness of the method discussed in this section, we assume $f = -1$ in Eq. (7.46) and the following boundary conditions on a square domain of unit length:

$$u(x,0) = u(0,y) = 0 \tag{7.53}$$

$$\frac{\partial u}{\partial x}\Big|_{(1,y)} = \frac{\partial u}{\partial y}\Big|_{(x,L)} = 0 \tag{7.54}$$

The exact solution to this boundary value problem is given by Eq. (7.32), and the plots of the true solution are shown in Fig. 7.8. The local approximations, T_i, are approximated by using the zeroth order ($m = 0$) and the first-order ($m = 1$) GLO-MAP weight functions. Further, to study the approximation error convergence with the order of basis functions, we use linear, quadratic, and cubic polynomials, orthogonal to the zeroth and the first-order weight functions, as basis functions. We also consider uniformly distributed points with different inter-nodal distances, h, to study the convergence and accuracy of the method. In all of the cases, the computed solution is tested on a total of 2,500 uniformly distributed points inside unit square.

    Figs. 7.11 and 7.12 show surface plots of the computed solution and its various partial derivatives using zeroth and first-order weight functions, respectively. From these figures, it is clear that both zeroth and first-order weight functions are able to approximate the solution accurately. However, as expected in case of the zeroth-order weight function, the various partial derivatives are discontinuous along the boundary of a particular sub-domain, Ω_s (see Fig. 7.13), while the first-order weight functions merge different local approximations to guarantee the continuity of solution and its various first derivatives. Fig. 7.14 shows the plot of relative error, e, with respect to nodal distance, h. As expected, the relative error decreases with a decrease in nodal point distance, h, and an increase in the order of basis functions. Further, comparing results of Fig. 7.14 with those of Fig. 7.9 reveals that in the case of the PUFEM algorithm, we achieve even better convergence (one order of magnitude). This is also due to the fact that even though the inter-nodal distance, h, is same for both of the algorithms, the local approximations in the case of the MLPG algorithm are computed using nodal points in a much larger domain to guarantee a well-conditioned linear system of equations for each local approximation.

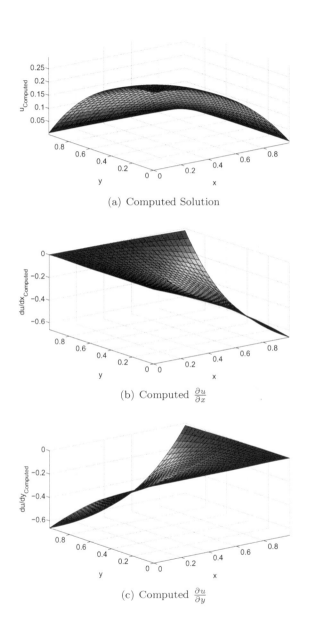

(a) Computed Solution

(b) Computed $\frac{\partial u}{\partial x}$

(c) Computed $\frac{\partial u}{\partial y}$

FIGURE 7.11

Computed Solution to the Poisson Equation Using the Zeroth-Order Weight Function.

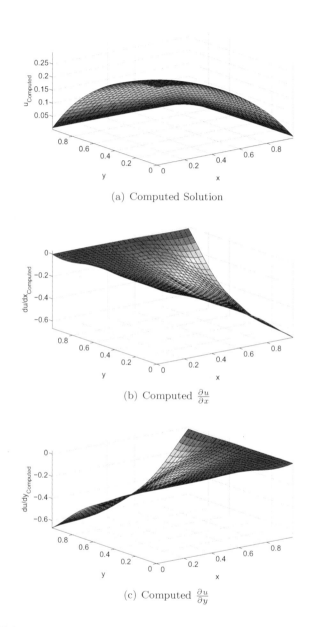

(a) Computed Solution

(b) Computed $\frac{\partial u}{\partial x}$

(c) Computed $\frac{\partial u}{\partial y}$

FIGURE 7.12

Computed Solution to the Poisson Equation Using the First-Order Weight Function.

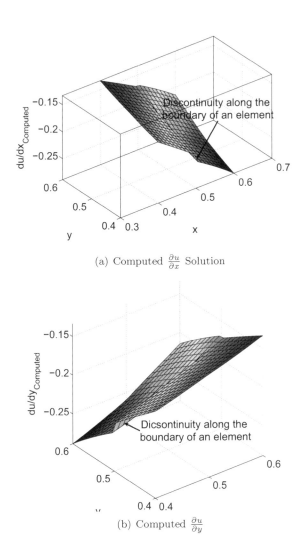

(a) Computed $\frac{\partial u}{\partial x}$ Solution

(b) Computed $\frac{\partial u}{\partial y}$

FIGURE 7.13
Computed Solution to the Poisson Equation Using the Zeroth-Order Weight
Function.

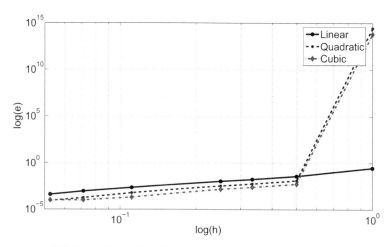

(a) Relative Error Plot Using the Zeroth-Order Weight Function

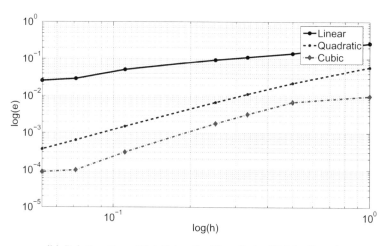

(b) Relative Error Plot Using the First-Order Weight Function

FIGURE 7.14
Relative Error Plot for the PUFEM Algorithm.

7.3.2 Fokker-Planck-Kolmogorov Equation

The Fokker-Planck-Kolmogorov Equation (FPKE), or simply the Fokker-Planck Equation (FPE) [99], provides the exact description of the transition probability density function (pdf) for nonlinear dynamical systems and has applications in numerous fields of science and engineering [99, 100]. One may be interested in the determination of the response of engineering structures such as beams, plates, entire buildings, beams under random excitation (in structure mechanics [101]), or the propagation of initial condition uncertainty of an asteroid for the determination of its probability of collision with a planet (in astrodynamics [102]), or the motion of particles under the influence of stochastic force fields (in particle physics [103]), or simply the computation of the prediction step in the design of a Bayes filter (in filtering theory [104]). All of these applications require the study of the time evolution of the pdf, $p(t, \mathbf{x})$, corresponding to the state, $\mathbf{x}$, of the relevant dynamic system.

Analytical solutions exist only for stationary pdf and are restricted to a limited class of dynamical systems [99, 100]. Thus, researchers are looking actively at numerical approximations to solve the Fokker-Planck equation [98, 105–107], generally using the variational formulation of the problem. Several numerical and semi-numerical approaches, such as the global Galerkin method [106–108], finite differences [109], finite elements [110, 111], and multi-scale finite elements [112], have been developed to solve the FPKE. However, these methods are severely handicapped for higher dimensions because the generation of meshes for spaces beyond three dimensions is still impractical.

In this section, we illustrate the effectiveness of the PUFEM in solving the FPKE and to tackle the issue of excessive computation time. For the purpose of illustration, we shall restrict our concern to the stationary FPKE for dynamic systems of a certain form, namely, non-conservative nonlinear systems with a Hamiltonian-like energy function. The reason for this choice is that for such systems, a unique globally asymptotically stable stationary solution exists which can be determined analytically [100], and thus, numerical solutions can be compared against the analytical solution.

Let us consider a general n-dimensional noise-driven nonlinear dynamic system with uncertain initial conditions, given by the equation

$$\dot{\mathbf{x}} = \mathbf{f}(t, \mathbf{x}) + \mathbf{g}(t, \mathbf{x})\Gamma(t) \tag{7.55}$$

where $\Gamma(t)$ represents a Gaussian white noise process with the correlation function $\mathbf{Q}\delta(t_1 - t_2)$, and the initial state uncertainty is captured by the pdf $p(t_0, \mathbf{x})$. Then the time evolution of $p(t_0, \mathbf{x})$ is described by the following FPKE, which is a second-order pde in $p(t, \mathbf{x})$:

$$\frac{\partial}{\partial t}p(t, \mathbf{x}) = \mathcal{L}_{\mathcal{FP}}p(t, \mathbf{x}) \tag{7.56}$$

where

$$\mathcal{L}_{\mathcal{FP}} = \left[-\sum_{i=1}^{n} \frac{\partial}{\partial x_i} \mathrm{D}_i^{(1)}(.) + \sum_{i=1}^{n} \sum_{j=1}^{n} \frac{\partial^2}{\partial x_i \partial x_j} \mathrm{D}_{ij}^{(2)}(.) \right] \quad (7.57)$$

$$\mathrm{D}^{(1)}(t, \mathbf{x}) = \mathbf{f}(t, \mathbf{x}) + \frac{1}{2} \frac{\partial \mathbf{g}(t, \mathbf{x})}{\partial \mathbf{x}} Q \mathbf{g}(t, \mathbf{x}) \quad (7.58)$$

$$\mathrm{D}^{(2)}(t, \mathbf{x}) = \frac{1}{2} \mathbf{g}(t, \mathbf{x}) Q \mathbf{g}^{\mathrm{T}}(t, \mathbf{x}) \quad (7.59)$$

where $\mathcal{L}_{\mathcal{FP}}$ is the Fokker-Planck operator, $\mathrm{D}^{(1)}$ is the drift coefficient vector and $\mathrm{D}^{(2)}$ is the diffusion coefficient matrix. The drift vector captures the drifting apart of the mean of the propagated pdf from the propagated mean of the initial pdf. In the case where the underlying governing dynamics is deterministic, i.e., $g(t, \mathbf{x}) = 0$, and the source of uncertainty lies only in the initial state, the diffusion matrix is identically zero and the reduced FPE is called the Liouville equation.

Despite its innocuous appearance, the FPE is a formidable problem to solve because of the following issues:

1. *Positivity* of the pdf: $p(t, \mathbf{x}) \geq 0, \ \forall t \ \& \ x$

2. *Normalization* constraint of the pdf: $\int\limits_{-\infty}^{\infty} p(t, \mathbf{x}) \mathrm{d}V = 1$

3. No fixed *solution domain*: how to impose boundary conditions in a finite region and restrict numerical computation to regions where $p > 10^{-9}$

A local variational form for the stationary FPKE over the sub-domain, Ω_s, with boundary, Γ_s is

$$\int\limits_{\Omega_s} \mathcal{L}_{\mathcal{FP}}(\hat{p}(t, \mathbf{x})) v d\Omega + \alpha \int\limits_{\Gamma_s \cap \Gamma} \hat{p}(t, \mathbf{x}) v d\Gamma = 0 \quad (7.60)$$

where v is a test function and α is a penalty parameter used to enforce the boundary condition $p(t, \infty) = 0$. The test functions in the above variational equation have compact support on Ω_s. Note that the integral is carried out over the local domain and the boundary condition is enforced over the part of the local boundary that coincides with the global boundary. As discussed earlier, after some algebraic manipulations, we get the following system of algebraic equations involving stiffness matrix $\mathbf{K}$:

$$\mathbf{K}\mathbf{a} = 0 \quad (7.61)$$

with

$$K_{ij} = \int\limits_{\Omega_s} \mathcal{L}_{\mathcal{FP}}(f_j(\mathbf{x}) \phi^T(\mathbf{x})) v_i d\Omega + \alpha \int\limits_{\Gamma_s \cap \Gamma} f_j(\mathbf{x}) \phi(\mathbf{x}) v_i d\Gamma \quad (7.62)$$

where $v_i = f_i\phi$ is the test function associated with the ith node. The implementation of the PUFEM can be carried out as discussed in the previous section. Clearly, for this problem the solution lies in the span of the null-space of the matrix $\mathbf{K}$. Theoretically, the null-space of $\mathbf{K}$ is unique if there exists a globally asymptotically stable stationary solution of the FPKE; however, this may not be true in the numerical implementation. We mention that if the parameter α is chosen to be too large, it may cause null space of $\mathbf{K}$ with dimension greater than 1. In such a case, one may study the equation error to determine the best solution. Alternatively, the penalty parameter, α, can be tuned to obtain a 1-D null space, or the boundary conditions can be implemented not as $\hat{p}(t, \infty) = 0$, but as a very small value, $\hat{p}(t, \infty) = \epsilon$ ($\sim 10^{-9}$), so that the RHS of Eq. (7.61) is not zero. The latter approach gives highly acceptable results, even with a very coarse tuning of α [105]. These issues are discussed in much more detail in Ref. [105].

For numerical simulation purposes, we consider the FPKE corresponding to the following 2-D nonlinear dynamic system:

$$\ddot{x} + \eta\dot{x} + \alpha x + \beta x^3 = g(t)\mathcal{G}(t) \qquad (7.63)$$

Eq. (7.63) represents a noise-driven Duffing oscillator with a soft spring ($\alpha\beta < 0, \eta > 0$), and includes damping to ensure the presence of a stationary solution. For simulation purposes, we use $\alpha = -15$, $\beta = 30$ and $\eta = 10$. The functional form of the true stationary solution, $p_s(x, \dot{x})$, is [100]

$$p_s(x, \dot{x})\mid_{true} = \mathcal{C} \exp\left(-2\frac{\eta}{g^2 Q}\left(\frac{\alpha x^2}{2} + \frac{\beta x^4}{4} + \frac{\dot{x}^2}{2}\right)\right) \qquad (7.64)$$

where $\mathcal{C}$ is a normalization constant. Note that the stationary pdf is an exponential function of the steady-state system Hamiltonian, scaled by the parameter $-2\frac{\eta}{g^2 Q}$ [100] and is bimodal in nature, as shown in Fig. 7.15(a).

The expressions for the elements of the stiffness matrix and load vector are given by

$$K_{ij} = \int_{\Omega_s}\left\{-\dot{x}\frac{\partial}{\partial x} + (\eta\dot{x} + \alpha x + \beta x^3)\frac{\partial}{\partial \dot{x}} + \eta + \frac{g^2 Q}{2}\frac{\partial^2}{\partial \dot{x}^2}\right\}[f_j(\mathbf{x})\phi^T(\mathbf{x})]v_i d\Omega +$$

$$\alpha\int_{\Gamma_s \cap \Gamma} f_j(\mathbf{x})\phi^T(\mathbf{x})v_i d\Gamma \qquad (7.65)$$

$$f_j = \alpha\epsilon\int_{\Gamma_s \cap \Gamma} v_i d\Gamma \qquad (7.66)$$

With regard to the discussion of rank deficiency of $\mathbf{K}$, the boundary condition was implemented as $\hat{p}(t, \infty) = \epsilon$ ($\approx 10^{-9}$), resulting in a non-zero load vector. Fig. 7.15(b) shows the solution obtained using the PUFEM algorithm on a 14×14 rectangular grid. For the sake of comparison, the above problem was

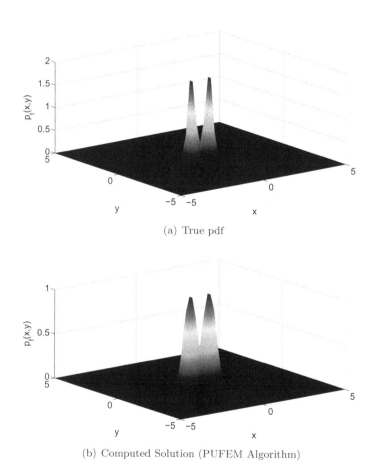

(a) True pdf

(b) Computed Solution (PUFEM Algorithm)

FIGURE 7.15
Numerical Results Using the PUFEM Algorithm.

also solved using the MLPG method. However, the convergence characteristics of the PUFEM are found to be significantly better than the MLPG method, as seen in Fig. 7.16. Although the convergence rate of the latter algorithm is faster, the RMS-error values are higher. The fast rate of convergence of the MLPG is most likely due to the decrease in interpolation errors as the density of nodes is increased. Furthermore, the PUFEM algorithm is considerably more computationally efficient, i.e., the time of execution of the PUFEM algorithm is much less than the computation time for MLPG. This is primarily due to the fact that MLPG requires the solution to several MLS problems in the process of evaluating the weak form integrals. Thus, for this particular problem, the PUFEM provides improvement in accuracy and efficiency over other meshless methods based on MLS.

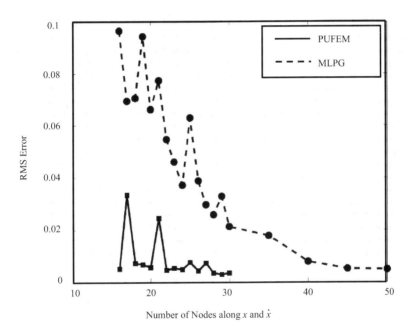

FIGURE 7.16

Comparative Convergence Characteristics of PUFEM and MLPG for Solving FPKE Corresponding to the Damped Duffing Oscillator.

7.4   Summary

This chapter discusses two different meshless algorithms to solve PDEs in an efficient manner. Both of the algorithms impose essential boundary conditions using penalty terms. Both of the algorithms also allow the inclusion of a priori knowledge about the solution of the PDE at hand by selecting appropriate basis functions for each local approximation. The performance of these algorithms can be evaluated by considering the following criteria:

- **Shape function selection:** The PUFEM offers great flexibility in the selection of local approximation spaces as it is possible to introduce different number of independently chosen basis functions in the different sub-domains. The shape functions are constructed by simply multiplying the basis functions with the PU pasting functions. In most other meshless methods, the shape functions are constructed using data fitting algorithms like the MLS. Consequently, while it is possible to use non-polynomial functions in the approximation space, it is a relatively difficult task to use different basis functions in different regions of the solution domain. Conventional FEM typically uses only polynomial shape functions in the approximation space, the order of which is determined by the shape of the finite element. Finally, there is no scope for local error improvement of the solution in the global methods because of the nature of the formulation.

- **Error characteristics:** The convergence of the PUFEM is expected to be superior to that of FEM, especially with the use of special functions in the approximation space which directly improves its approximability. From our experience in the current application, we conclude that the convergence characteristics of PUFEM are better than those of MLPG, using the same basis function set [98, 105]. This could be partially due to the absence of interpolation errors in the PUFEM, whereas in MLPG, additional errors are introduced due to interpolation required to find the solution at points other than the solution nodes.

- **Computational load:** If we divide the computational load of the various algorithms into three stages − pre-processing (grid generation), integration (evaluation of the weak form integrals) and post-processing (finding a solution at several points in the domain) − we get the following relative ordering: the PUFEM and MLPG rank above FEM in the pre-processing stage because the required grid information is minimal in the former. PUFEM and FEM are faster than most other meshless methods in the integration stage because the latter requires the solution to an MLS problem for every quadrature point used for the numerical evaluation of the integrals. The PUFEM ranks above both FEM and other meshless methods in the post-processing stage because it inherently provides a functional form of the approximation, i.e., no "new" interpolation procedure is required to construct the solution at any given point in the domain, as is true in the other methods. However, this very aspect (functional form of the approximation inside each sub-domain) implies that the PUFEM demands greater computational memory than other meshless methods, which typically give only the solution values at the nodes. Clearly, the functional form requires a greater number of parameters to be stored for solution characterization, as opposed to specifying only the solution value at the various nodes.

- **Applications to high-dimensional problems**: The PUFEM and MLPG stand out from the conventional FEM in the aspect of implementation to higher-dimensional problems because mesh generation in three and higher dimensions is still not practical. Comparing MLPG and the PUFEM in this respect, the PUFEM has some advantage because of its simpler algorithm structure and smaller time of execution. While MLPG requires computational time in the face of the curse of dimensionality, it provides benefit in the aspect of memory requirement because of the smaller number of degrees of freedom involved in the approximation; this advantage increases as dimension of the space increases.

In summary, the PUFEM emerges as the algorithm that is the easiest to implement in the applications considered because of its simplicity. While it involves only the numerical evaluation of integrals in the weak form, followed by the inversion of the stiffness matrix, most other meshless methods require the solution to several MLS problems in addition to these tasks. Furthermore, if polynomial basis functions are used, the PUFEM integrals can be found analytically. On the other hand, most other meshless methods require numerical integration even with polynomial bases, because the shape functions resulting from the MLS procedure are not polynomials. Both algorithms are equally easy to extend to higher dimensions. MLPG has an advantage over the PUFEM in one critical aspect. It can be easily used on domains with boundaries of arbitrary shape, whereas in the current established PUFEM framework, it is essential to have the solution domain as an N-hypercube. Finally, we mention that both of the algorithms can, in principle, be extended to handle high-dimensioned partial differential equations. All three approaches will be affected to a yet-to-be established degree by the "curse of dimensionality," when very high-dimensioned problems are addressed.

8

Control Distribution for Over-Actuated Systems

Glory is fleeting, but obscurity is forever.

Napoleon Bonaparte

8.1 Introduction

In this chapter, we consider the control distribution problem for highly over-actuated systems. In the case of over-actuated systems, there is redundancy in the total number of actuators, compared to the number of equations of motion that govern the dynamics of the system. For example, in the F-16 aircraft, two-axis thrust vectoring is used along with conventional control surfaces (ailerons, rudder, and elevator) and the throttle, resulting in seven control inputs to produce six net control forces and moment components needed to control the six degrees of freedom. As a second example, consider the attitude control of a satellite. Frequently there will be two sets of actuators: one set consists of twelve on-off small jets or thrusters configured in six pairs such that one pair can cause a positive pitch rotation, and another pair can cause a negative pitch rotation, and so on for the other three axes of rotation. The firing direction of the thrusters can be changed so that each pair can also cause pure translations in each of the three directions. This is precisely the pattern used for translational and pointing control for the space shuttle. These on-off thrusters result in bang-bang control and are not suitable for fine pointing. For many spacecraft, a second set of actuators, namely momentum exchange devices (such as control moment gyros), are incorporated into the system design to allow smooth, variable-amplitude torques for precision maneuvers. A minimum of three control moment gyros are required to generate a torque in any commanded direction, but normally four or more are used to allow redundancy, increased torque capability, and to avoid certain singular conditions that arise for such momentum transfer devices [77]. The space station utilizes four control moment gyros for the primary attitude control system. The combination of the twelve thrusters and four momentum exchange devices provides a significant level of redundancy to increase the frequency bandwidth and allow flexible budgeting of fuel and electricity as well as lead to increased reliability. Generally, control distribution approaches exploit the set of actuation possibilities using some "best actuation commands" to achieve the desired system state change, while taking into account constraints on individual actuators, power, and fuel reserves, avoiding undesirable actuation configurations or singularities, and similar considerations. The current state of the practice has been successfully targeted on specific physical systems with fewer than twenty actuators for up to six degrees of freedom. The present chapter seeks to generalize these conventional ideas so that massively redundant actuation can be accommodated for the high-dimensioned systems anticipated for the future.

In the last decade there have been significant advances in the fields of Micro Electro Mechanical Systems (MEMS), Nano Electro Mechanical Systems (NEMS), and nano bio systems. It is anticipated that advances in these technologies will lead to the development of adaptive, intelligent and shape-controllable structures for future aircraft and space systems. The design of such advanced systems involves control of the shape of the structures, and, in contrast to the slightly over-actuated F-16, will involve highly redundant micro- and nano- level manipulations (actuation). Actuators embedded in conventional structural materials at discrete or distributed locations can be used to achieve these objectives by changing ("morphing") surface shape. Currently existing smart structure actuators are Shape Memory Alloys (SMAs), piezoelectric and electrostrictive ceramics, electro- and magneto-rheological fluids, Synthetic Jet Actuators (SJAs) and active elastomers. Current research activities in nanotechnologies are aimed at engineering these functionalities into materials at molecular and atomic scales. Such systems can conceivably have quite a large number of actuators ($\sim 10^6$), which could collectively produce the required control effort, but lead to controller design problems in which the number of control components may exceed the required number of degrees of freedom of the system by several orders of magnitude. There is no doubt that with such massive redundancy in control variables, one may, in principle, achieve new levels of fault-tolerant control. However, the main control theory and computational challenge lies in developing control approaches that scale efficiently with a large number of control variables. Among the many practical challenges associated with the design of redundant control variables are:

1. *Actuator Models*: The issue at hand is to derive comprehensive mathematical models that capture the input-output behavior of these actuators so that one can derive automatic control laws to command desired shape and behavior changes. This mapping should also generate an envelope that bounds the maximum reachable control inputs.

2. *Dimensionality*: Dealing with a large number of actuators vs. a smaller numbers of Degrees of Freedom (DoF).

3. *Numerical Conditioning*: Solving for a large number of control inputs ($\sim 10^6$) using conventional methods generally leads to an ill-conditioned numerical problem.

4. *Computational Cost*: The controller design must be compatible with near-real-time computing, as ultimately will be required.

5. *Sensing and communication requirements.*

In Chapters 3 and 5, non-parametric, multi-resolution, adaptive input-output modeling approaches were discussed to capture macro static and dynamic models directly from experiments which can be used, in principle, for these embedded systems. In Ref. [70], we used the Radial Basis Function (RBF)-based non-parametric mathematical model to learn the mapping between the SJA parameters (synthetic jet frequency, amplitude, direction, etc. for each of several SJA), and the resulting aerodynamic lift, drag, and moment. In this chapter, our main interest is to present a general control distribution technique that can be applied for very large scale dynamical systems. This chapter has three main objectives. The first and most important objective is to present a recursive control distribution approach using adaptive distribution functions to address the issue of dimensionality and computational efficiency. The second objective of this chapter is to establish insight on the implementation of the recursive algorithm and learning the parameters of the distribution functions. The third and final objective of this chapter is to compare the new algorithm with existing algorithms in terms of computational efficiency and distribution accuracy. This should provide insights on the relative merits of several approaches.

The structure of the chapter is as follows: First, a problem statement for the control distribution problem is introduced, followed by a brief review of some existing control distribution algorithms. Then, a novel recursive control distribution algorithm is introduced, and finally, the new algorithm is validated and compared by simulating various test cases.

8.2 Problem Statement and Background

Let us consider a general dynamical system governed by the following differential and algebraic equations:

$$\dot{\mathbf{x}} = f(\mathbf{x}, t) + g(\mathbf{x})\mathbf{u} \tag{8.1}$$

$$\mathbf{y} = h(\mathbf{x}, \mathbf{u}) \tag{8.2}$$

where $\mathbf{x} \in \mathcal{R}^n$ is the dynamical system state vector, $\mathbf{u} \in \mathcal{R}^p$ is a vector of actuator inputs, and $\mathbf{y} \in \mathcal{R}^m$ is a vector of system outputs measured by various sensors on board. The trajectory-tracking control problem for the dynamical system given by Eqs. (8.1) and (8.2) is defined as follows: *Given a model of system dynamics and the desired smooth state trajectory, $\mathbf{x}_d$, compute the appropriate control vector that will drive the system along the desired state trajectory.* Depending upon the relative size of the system state, actuator input vectors, and the controllability of the system, there are three possible outcomes to the control problem:

1. Infinite number of solutions $(p > n)$: pick the best one

2. One unique solution $(p = n)$

3. No solution $(p < n)$: find best approximate solution

In this chapter, we are interested in the first case when $p > n$, i.e., over-actuated systems. Conventional linear and nonlinear control methodologies are applicable for only the second and third cases when $p \leq n$ and the control problem is, at least in some respects, more complicated than for over-actuated systems. While redundant actuators frequently afford us the opportunity to avoid the lack of controllability issues that arise for some under-actuated cases, we encounter other challenges associated with increased dimensionality of the control vector. To make use of recent advances in the control literature, the control problem for the over-actuated system generally is divided into two parts:

1. First, conventional control laws are designed specifying the instantaneous total physical control (e.g., resultant forces and moments, also known more generally as *virtual control variable*). Equations of motion, dictated by Newton's second law, are used together with appropriate control theory to derive these control laws, and the total number of virtual control variables is frequently equal to the number of mechanical degree of freedoms.

2. In the second step, the total control effort is distributed among individual actuators, taking into account various actuator constraints. Algebraic or dynamic models describing the relationship between actuator input and output are used, along with optimality criteria, to find the set of actuator inputs that produce the resultant net virtual control.

To illustrate the procedure, let us consider the example of the control of an advanced morphing aircraft with thousands of actuators embedded in the aircraft wing. It is a well-known fact that the gross "rigid body" aircraft dynamics represents a 6 DoF motion, and theoretically one requires 3 resultant forces and 3 moments along yaw, pitch, and roll axes to control the aircraft. In the first step, these three forces and moments constitute the 6-dimensional virtual control vector, $\mathbf{v}$, which can be solved by using conventional control methodologies and the equations of motion governed by Newton's second law

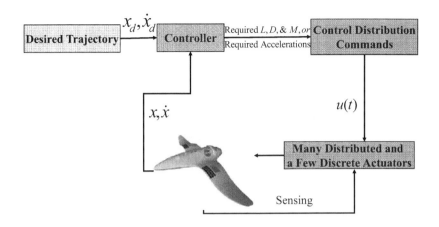

FIGURE 8.1
Control of Highly Over-Actuated System.

of motion. In the second step, the desired virtual control vector, $\mathbf{v}$, is allocated among individual actuators to find actual control variables like voltage input to each actuator; this approach assumes that the virtual control vector is physically possible, given the constraints on individual actuator inputs. This two-step process for control of over-actuated systems is shown in Fig. 8.1. According to this two-step control methodology, the control distribution problem can be defined as follows:

Given the desired profile, or instantaneous value for virtual/physical control vector $\mathbf{v}(t) \in \mathcal{R}^q$, find the true input vector $\mathbf{u}(t) \in \mathcal{R}^p$ such that the following is true:

$$\mathbf{g}(\mathbf{u}(t)) = \mathbf{v}(t) \tag{8.3}$$

$$\underline{\mathbf{u}} \leq \mathbf{u} \leq \bar{\mathbf{u}} \tag{8.4}$$

$$\underline{\dot{\mathbf{u}}} \leq \dot{\mathbf{u}} \leq \bar{\dot{\mathbf{u}}} \tag{8.5}$$

where $\mathbf{g}(\mathbf{u}(t))$ represents the transformation from higher dimension space, $\mathcal{R}^p$, to lower dimension space, $\mathcal{R}^q$. Further, Eqs. (8.4) and (8.5) describe the position and rate constraint on actuator response. It should be mentioned that for successful design of the control distribution algorithm, the accurate knowledge of $\mathbf{g}(.)$ is very essential. The input-output algorithms presented in Chapters 3 and 5 can be used to learn the input-output mapping for various kinds of actuators. In Ref. [70], we use the DCG algorithm to learn the input-output mapping of a synthetic jet actuator directly from experiments. Generally, a linear model is desired between virtual control vector, $\mathbf{v}$, and true control vector, $\mathbf{u}$, such that Eq. (8.3) can be replaced by the following equation:

$$\mathbf{B}\mathbf{u}(t) = \mathbf{v}(t) \qquad (8.6)$$

where $\mathbf{B} \in \mathcal{R}^{q \times p}$ is a matrix with rank q. In the case of the linear control distribution problem, the solution lies on the intersection of the hyper-surface, $\mathbf{B}\mathbf{u} = \mathbf{v}$, and position control hyper-box defined by Eqs. (8.4) and (8.5). We mention that Eqs. (8.3) and (8.6) are written with the assumption that actuator response is instantaneous, i.e., actuator dynamics is negligible, which will not generally be true. For example, as discussed in Ref. [113], the actuator dynamics is certainly present in the case of SJAs. However, this assumption is valid to a degree of approximation, if the closed-loop system is designed to be substantially slower than the actuator dynamics.

In Ref [114], numerous advantages that ensue from breaking the control problem into two steps are discussed in detail. Practically, dividing the control problem into two steps allows us to exploit individual actuators to their full level without degrading the closed-loop performance of the controller design in the first step. Further, actuator constraints, saturation and failure can be handled more efficiently. If one actuator saturates and fails, then another actuator may be used to make up the difference. In other words, the reconfiguration of different actuators can be performed in the event of an actuator failure, without redesigning the control law in the first step. Another main advantage is that actuator utilization can be optimized independently for a specific application. For example, thrust vectoring can be used as auxiliary control to obtain high maneuverability. Similarly, the use of trailing edge SJAs is preferable at low Angle of Attack (AOA), while the leading edge SJAs are useful in the case of flow separation.

Many algorithms [114, 115] have been suggested in the literature to solve the control distribution problem for over-actuated systems. The generalized inverse, known as the minimum norm pseudo-inverse, is a frequently used method to compute the solution for control distribution problems. For the over-actuated case, the generalized inverse solution is obtained by minimizing the 2-norm of the true control vector, $\mathbf{u}$, subject to the constraint given by Eq. (8.6). In the absence of any constraints on control variable $\mathbf{u}$, the explicit generalized inverse solution is given by the following equation:

$$\mathbf{u} = \mathbf{B}^T (\mathbf{B}\mathbf{B}^T)^{-1} \mathbf{v} \qquad (8.7)$$

A more computationally robust minimum norm solution can be computed using SVD algorithms. In a more general approach, the following optimization problem is defined to solve the control distribution problem:

$$Minimize : \ \|\mathbf{u}\|_p \qquad (8.8)$$

subject to

$$\mathbf{B}\mathbf{u} = \mathbf{v} \qquad (8.9)$$

$$\underline{\mathbf{u}} \leq \mathbf{u} \leq \bar{\mathbf{u}} \qquad (8.10)$$

$$\underline{\dot{\mathbf{u}}} \leq \dot{\mathbf{u}} \leq \bar{\dot{\mathbf{u}}} \qquad (8.11)$$

where $\|\mathbf{u}\|_p$ is the p norm of vector $\mathbf{u}$ defined as follows:

$$\|\mathbf{u}\|_p = \left(\sum_{i=1}^{p} |u_i|^p \right)^{\frac{1}{p}} \tag{8.12}$$

Most commonly used norms are 1- and 2-norm. Although a variety of norm definitions can be used to solve the control distribution problem, the use of different norms leads to different numerical methods to solve the problem, and the suitability of a given norm will be dictated by the viability of the algorithm and the physical characteristics of the resulting control distribution solution. The use of 2-norm leads to a conventional quadratic optimization, problem whereas the use of 1-norm or ∞-norm leads to the linear programming problem. Generally, active set methods [116, 117], primal and dual simplex methods [118, 119] and interior point optimization methods [120] are used to solve linear and quadratic optimization problems. Active set methods solve an optimization problem by partitioning inequality constraints into two sets: active and inactive. The inactive constraints are ignored while solving the problem, whereas the active set constitutes the working set for the solution at any given step. Then, the problem is solved by moving on the surface defined by the working set. These methods search for a solution along the edges and faces of the feasible set by solving a sequence of equality-constrained quadratic programming problems. On the other hand, the conventional simplex method solves the optimization problem by searching from vertex to vertex on the boundary of the feasible polyhedron, repeatedly increasing the objective function until either an optimal solution is found, or it is established that no solution exists. In principle, the time required might be an exponential function of the number of variables, and this can happen in some contrived cases.

To deal with the problems associated with the simplex method, interior point optimization methods are used. Unlike the simplex method, interior point optimization methods do not search for the solution from vertex to vertex, but search only through the interior of the feasible region. In brief, active set and interior point methods differ from the simplex method in that the solution in this case need not be on vertices of the feasible set. Generally, active set methods are used to solve the quadratic optimization problem, while the simplex method is mainly used to solve linear programming problems. Quadratic optimization problems are more difficult to solve than linear programming problems because unlike the solution to linear programming problems, the quadratic problem solution may use all variables of the problem. Therefore, the 2-norm solution tries to distribute the total control effort among all of the control inputs, whereas the 1-norm solution utilizes as few control variables as possible, and may lead to saturation. Finally, we mention that the percentage of attainable control effort using the optimization problem solution can be quite small, depending on a number of factors, including the number of control variables and the definition of the norm used in the optimization problem [121]. Also, most of the available numerical methods can handle an unlimited number of variables and constraints, subject to the availability of computer time, memory and numerical conditioning of the particular application. Practical experience tells us, however, that solving a large-scale optimization problem is not desirable for real-time problems.

For very large-scale systems, the use of a hierarchical approach, known as *daisy chain* [122], is discussed. The daisy chain method uses heuristic logic to divide the p control inputs into P groups, $\{\mathbf{u}^1, \mathbf{u}^2, \cdots, \mathbf{u}^P\}$. Initially, the control variable, in the first ("primary") group, $\mathbf{u}^1$, are used, i.e., $\mathbf{B}^1\mathbf{u}^1 = \mathbf{v}$. If the primary control variables in $\mathbf{u}^1$ satisfy all the constraints of Eqs. (8.4)−(8.6), then the distribution is successful. Otherwise, the control variables in the primary group $\mathbf{u}^1$ which violate constraints of Eqs. (8.4) and (8.5) are saturated, and control variables in group $\mathbf{u}^2$ are used to provide the rest of the control effort, i.e., control effort equal to the difference between the total control effort, $\mathbf{v}$, and the control effort produced by control variables in group $\mathbf{u}^1$. This procedure is recursively repeated until the desired control effort is produced or all control variables are used. The main advantage of the daisy chain method is that primary and secondary actuators can be expected to be used most frequently, and the higher control groups are used only when necessary. On the other hand, the main disadvantage of this approach is that this procedure does not take into account the actuator constraints directly, and employs the simple heuristic of saturating the actuators, where they are commanded more control effort than their physical limit. Generally, saturation of actuators is not desirable for many problems. Further, for a very large-scale system this approach is not desirable, as the number of control variables in each group and the number of groups can be quite large for such systems and make this algorithm computationally inefficient. Finally, the ordering of primary and secondary control groups is heuristic, and the results obtained may be far from optimal in particular applications.

In Refs. [115,123], different approaches to establish hierarchical algorithms are explored for control allocation in a large-scale distributed system. The hybrid algorithm approach is based upon a *divide and conquer* method and works by breaking a high-dimensional problem into a number of smaller problems that can either be reduced in size, or solved using optimization algorithms discussed earlier in this section. The optimization algorithms can range from discrete optimal search to continuous constrained optimization problems. The method discussed in Ref. [123] is tested for the control distribution problem among hundreds of actuators to control the translation of a sheet of paper over an air-jet table. The optimal algorithm yields low errors, but the time required to compute the solution increases exponentially with the number of actuators because standard optimization algorithms are used to find the discrete control variables at various scales. Therefore, even though the hierarchical algorithm discussed in Refs. [115,123] is shown to work for reasonably large problems, they are not computationally feasible for very large-scale problems where the number of control variables can be in the millions. For example, let us consider a problem of control distribution among 10^6 actuators. In this case, even though one divides this highly redundant actuation problem into 1,000 small problems, then each small problem also has 1,000 optimization variables to be solved for, which can be computationally infeasible depending upon the computing resources and the frequency with which the solution must be updated. Here, we propose the use of *distribution functions* to reduce the number of control variables by an order of magnitude. The main idea is to approximate the feasible solution set by making use of continuous functions. These functions are defined by a few parameters, and spatially distribute the controls by interpolating the inputs to each discrete actuator. However, if one uses a global set of distribution functions, then the number of distribution functions can be very high in order to have a reasonable approximation of the feasible set. Therefore, a hierarchical approach is used to improve the accuracy of the feasible set only locally as required, and to deal with the issue of high dimensionality. Qualitatively, the distribution function approach is motivated by the anticipation that spatially dense actuation will usually be locally correlated in space and time, especially as the continuous actuation limit is approached.

In this chapter, our main focus is to design an efficient control distribution algorithm to generate commands to a highly redundant system in real-time. To deal with the issues of high dimensionality, a hierarchical approach is proposed which makes use of specially designed distribution functions for control distribution purposes. In the rest of this chapter, the concept of distribution functions and the detailed description of the proposed hierarchical algorithm are discussed.

8.3   Control Distribution Functions

In the previous section, a brief introduction to various control distribution algorithms is given. Theoretically, each algorithm has the ability to handle an unlimited number of control variables subject to the availability of time, processing power and memory. However, in the case of a very large-scale distribution problem, most of these algorithms fail to compute the solution given practical limitations on processing power, time and memory. Basically, the main problem is the lack of a tool to reduce the dimensionality of the problem to the desired order. We introduce the idea of distribution functions to approximate the control effort so that the given problem can be solved with a modest computational burden.

Let us consider a general problem of distributing virtual/physical control vector $\mathbf{v}(t) \in \mathcal{R}^q$ among true input vector $\mathbf{u}(t) \in \mathcal{R}^p$ such that the following is true:

$$\mathbf{Bu}(t) = \mathbf{v}(t) \tag{8.13}$$

$$\underline{\mathbf{u}} \leq \mathbf{u} \leq \bar{\mathbf{u}} \tag{8.14}$$

$$\underline{\dot{\mathbf{u}}} \leq \dot{\mathbf{u}} \leq \bar{\dot{\mathbf{u}}} \tag{8.15}$$

where $\mathbf{v}(t)$ denotes the relevant physical forces which depend on actual displacement and velocity error vectors, and $\mathbf{u}(t)$ is a vector of actual control inputs. As mentioned earlier, we neglect the actuator dynamics, assuming that actuator dynamics is much faster than the closed-loop dynamics of the system under consideration. Further, rate constraints given by Eq. (8.15) can be converted into position constraints by approximating $\dot{\mathbf{u}}(t)$ by a first-order finite difference approach:

$$\underline{\dot{\mathbf{u}}} \leq \frac{\mathbf{u}(t) - \mathbf{u}(t^-)}{\Delta t} \leq \bar{\dot{\mathbf{u}}} \tag{8.16}$$

$$\underline{\dot{\mathbf{u}}}\Delta t + \mathbf{u}(t^-) \leq \mathbf{u}(t) \leq \bar{\dot{\mathbf{u}}}\Delta t + \mathbf{u}(t^-) \tag{8.17}$$

where $t^- = t - \Delta t$. As a consequence of this, the control distribution problem can be considered with a constraint on actuator response $\mathbf{u}(t)$ only.

$$\mathbf{u}_l(t) \leq \mathbf{u}(t) \leq \mathbf{u}_u(t) \tag{8.18}$$

where

$$\mathbf{u}_l(t) = \min\left(\underline{\mathbf{u}}, \underline{\dot{\mathbf{u}}}\Delta t + \mathbf{u}(t_1)\right) \tag{8.19}$$

$$\mathbf{u}_u(t) = \max\left(\bar{\mathbf{u}}, \bar{\dot{\mathbf{u}}}\Delta t + \mathbf{u}(t_1)\right) \tag{8.20}$$

Now, the optimization problem to solve the control distribution problem can be redefined as follows:

$$\min_{\Omega(t)} : \|\mathbf{u}\|_p \tag{8.21}$$

subject to

$$\mathbf{Bu}(t) = \mathbf{v}(t) \tag{8.22}$$

$$\mathbf{u}_l(t) \leq \mathbf{u}(t) \leq \mathbf{u}_u(t) \tag{8.23}$$

where $\Omega(t)$ denotes the set of feasible solutions and depends upon the total physical/virtual control effort, $\mathbf{v}(t)$. Now, assume that all actuators are spatially distributed over some surface, and there exists a set of distribution functions $\boldsymbol{\phi}(\mathbf{x}) = \{\phi_i(\mathbf{x})\}$ such that

$$span(\boldsymbol{\phi}(\mathbf{x})) = \Omega(t) \tag{8.24}$$

where $\mathbf{x} \in \mathcal{R}^M$ is a vector of spatial coordinates. We mention that generally M is equal to 2 or 3. As a consequence of Eq. (8.24), any feasible solution

$\mathbf{u}(t)$ can be approximated as a linear combination of distribution functions $\phi_i(\mathbf{x})$.

$$\mathbf{u}(\mathbf{x}, t) = \sum_{i=1}^{N} a_i(t)\phi_i(\mathbf{x}) \tag{8.25}$$

where $\mathbf{a}(t) \in \mathcal{R}^N$ is a vector consisting of amplitudes of various distribution functions $\phi_i \in \mathcal{R}^N$. Now, the true control vector, $\mathbf{u}$, can be written as

$$\mathbf{u} = \mathbf{\Phi}(\mathbf{x})\mathbf{a} \tag{8.26}$$

where

$$\mathbf{\Phi} = \begin{bmatrix} \phi_1(\mathbf{x}_1) & \phi_2(\mathbf{x}_1) & \cdots & \phi_N(\mathbf{x}_1) \\ \phi_1(\mathbf{x}_2) & \phi_2(\mathbf{x}_2) & \cdots & \phi_N(\mathbf{x}_2) \\ \vdots & & \ddots & \vdots \\ \phi_1(\mathbf{x}_p) & \phi_2(\mathbf{x}_p) & \cdots & \phi_N(\mathbf{x}_p) \end{bmatrix} \tag{8.27}$$

and, $\mathbf{x}_i$ denotes the spatial coordinates for the ith control variable, u_i.

Further, substituting for $\mathbf{u}(t)$ from Eq. (8.26) in Eqs. (8.21)–(8.23), we get the following optimization problem for the amplitude vector, $\mathbf{a}$,

$$\min : \|\mathbf{\Phi}(\mathbf{x})\mathbf{a}\|_p \tag{8.28}$$

subject to

$$\mathbf{B}\mathbf{\Phi}(\mathbf{x})\mathbf{a} = \mathbf{H}(\mathbf{x})\mathbf{a} = \mathbf{v} \tag{8.29}$$

$$\mathbf{u}_l \leq \mathbf{u} \leq \mathbf{u}_u \tag{8.30}$$

It should be noted that in this case one needs to solve for the N-dimensional amplitude vector, $\mathbf{a}(t)$, as compared to the p-dimensional true control variable vector, $\mathbf{u}(t)$, in the case of the optimization problem described by Eqs. (8.21)–(8.23). So, the reduction in the dimensionality of the problem depends upon the relative values of p and N. A key question regarding the selection of distribution functions, $\phi_i(.)$, is "How irregular is the feasible solution set $\Omega(t)$?" A globally valid set of distribution functions should be sufficient if $\Omega(t)$ is a well-connected set and all feasible solutions are globally smooth. However, if $\Omega(t)$ is a complicated set, or in the presence of high frequency local features in the feasible solution set, a more judicious selection of distribution functions will be required. While the brute force approach of using infinitely many basis functions is a theoretical possibility, it is intractable in a practical application because such an optimizer will have far too many parameters to determine and will not give any advantage in terms of dimensionality reduction. As discussed in Chapters 3 through 5, we consider Radial Basis Functions (RBFs) and global/local orthogonal polynomial basis functions as candidates for control distribution functions.

8.3.1   Radial Basis Functions

As discussed in Chapter 3, RBFs are basis functions whose response decreases monotonically with the increase in radial distance from their center location. The region of large value of each RBF can be confined to a local region around its center location, $\boldsymbol{\mu}$. Among many choices for radial basis functions, the Gaussian function is the most widely used because, among other reasons, the different parameters appearing in its description live in the space of inputs and have physical and heuristic interpretations that allow good starting estimates to be locally approximated. The most general Gaussian function can be written as

$$\phi_i(\|\mathbf{x} - \boldsymbol{\mu}_i\|, \boldsymbol{\sigma}_i, \mathbf{q}_i) = \exp\{-\frac{1}{2}(\mathbf{x} - \boldsymbol{\mu}_i)^T \mathbf{R}^{-1}(\boldsymbol{\sigma}_i, \mathbf{q}_i)(\mathbf{x} - \boldsymbol{\mu}_i)\} \qquad (8.31)$$

where $\boldsymbol{\mu}_i \in \mathcal{R}^M$ represents the center location of the Gaussian basis function, ϕ_i, whereas $\mathbf{R} \in \mathcal{R}^{M \times M}$ is a positive definite symmetric matrix which describes the shape and size of a Gaussian basis function. Now, substituting for $\phi_i(.)$ from Eq. (8.31) in Eq. (8.26), we get

$$\mathbf{u}(\mathbf{x}) = \sum_{i=1}^{N} \phi_i(\|\mathbf{x} - \boldsymbol{\mu}_i\|, \boldsymbol{\sigma}_i, \mathbf{q}_i)a_i \qquad (8.32)$$

Therefore, we need to solve for following parameters to use Gaussian basis functions as control distribution functions:

1. M parameters for the centers of the Gaussian function, i.e., $\boldsymbol{\mu}_i$

2. M parameters for the spread (shape) of the Gaussian function i.e., $\boldsymbol{\sigma}_i$

3. $\frac{n(n+1)}{2}$ parameters for rotation of the principal axis of the Gaussian function, i.e., $\mathbf{q}_i$

4. Amplitude a_i of the Gaussian function, ϕ_i

The main problem with the use of the Gaussian functions as control distribution functions is that, except for the amplitude vector, the various other parameters appear nonlinearly in Eq. (8.32) and necessitate the use of a nonlinear optimization algorithm to solve for their optimal value. The use of a nonlinear optimization algorithm may not be desirable for many practical reasons. To simplify the problem, one can pre-define "good choices" of the various parameters, except the amplitude vector, $\mathbf{a}$, whose optimal value can be found by solving the simpler algebraic optimization problem defined by Eqs. (8.28)−(8.30). The centers, $\boldsymbol{\mu}_i$, for various Gaussian basis functions can either be distributed uniformly over the input space, or they can be selected to make use of some a priori information about the grouping of actuators. Further, the spread parameter vector, $\boldsymbol{\sigma}_i$, can be chosen proportional to the shortest distance between $\boldsymbol{\mu}_i$ and the existing centers

$$\boldsymbol{\sigma}_i = \kappa \|\boldsymbol{\mu}_i - \boldsymbol{\mu}_{nearest}\| \qquad (8.33)$$

where κ is a user-defined parameter which accounts for the amount of overlap between different Gaussian functions. The rotation parameter, $\mathbf{q}_i$, can be assumed to be zero until some information is available on the control distribution surface. In the limiting case when $\kappa \to 0$, the Gaussian basis function approaches a dirac-delta function, and in this particular case, one would like to choose as many basis functions as the number of control variables, i.e., $p = N$. As a consequence of this, we return to the original optimization problem defined by Eqs. (8.28)–(8.30). Experience indicates that one would like to choose the parameter κ such that two neighboring basis functions overlap by at least 50%. Finally, one can iterate on the number of basis functions N depending upon whether a feasible solution exists or not. This provides a good compromise between "local dominance" and "trend sensing" of the RBF model. Initially, one can choose a small number of distribution functions with centers, $\boldsymbol{\mu}_i$, distributed uniformly in spatial coordinates, $\mathbf{x}_i$, and if a feasible solution is not found, then one can keep on increasing the number of basis functions until a feasible solution is obtained. The number N of basis functions should be initially set conservatively small (at least two orders of magnitude smaller than number of actuators). The outline of the control distribution algorithm using RBF functions is shown in Fig. 8.2. Note that the architecture is adaptive, and additional basis functions can be added. It is also easy to adaptively prune the representation, if the amplitude of local basis function becomes smaller than a tolerance.

8.3.2 Global Local Orthogonal Basis Functions

In Chapter 5, we introduced the idea of the Global Local Orthogonal MAPping (GLO-MAP) algorithm to approximate irregular surfaces. The same idea can be to interpolate an irregular control distribution surface, $\mathbf{u}(\mathbf{x}, t)$, with all the global and local advantages of the GLO-MAP approach to approximation. Introducing a set of grid points, $\{\bar{\mathbf{x}}_i\}_{i=1}^{Q}$, as approximation vertices having associated with weight functions, w_i, and local approximations, $\boldsymbol{\psi}_i(\mathbf{x}, \bar{\mathbf{x}}_i)$, we can approximate unknown control distribution surface, $\mathbf{u}(\mathbf{x}, t)$ as follows:

$$\mathbf{u}(\mathbf{x}, t) = \sum_{i=1}^{Q} w_i(\mathbf{x}, \bar{\mathbf{x}}_i) \boldsymbol{\psi}_i(\mathbf{x}, \bar{\mathbf{x}}_i) \tag{8.34}$$

where $w_i(.)$ represents a specially designed GLO-MAP weight function, and the ith local approximation, $\boldsymbol{\psi}_i(.)$, can be written as a linear combination of N polynomial basis functions, ϕ_{l_j}

$$\boldsymbol{\psi}_i(\mathbf{x}, \bar{\mathbf{x}}) = \sum_{j=1}^{N} a_{ij}(t) \phi_{l_j}(\mathbf{x}, \bar{\mathbf{x}}_i) = \boldsymbol{\phi}_l^T(\mathbf{x}, \bar{\mathbf{x}}_i) \mathbf{a}_i \tag{8.35}$$

where $\phi_{l_j}(.)$ can be chosen as the orthogonal polynomials of the GLO-MAP process. If particular insight is available on a specific problem, problem spe-

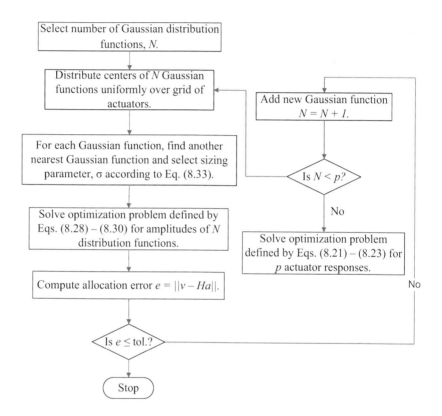

FIGURE 8.2

Flowchart for the Control Distribution Algorithm Using RBF Distribution
Functions.

cific distribution functions can also be introduced to enrich the distribution function set. Now, substitution of Eq. (8.35) in Eq. (8.34) leads to the following equation for the feasible control distribution surface, $\mathbf{u}(\mathbf{x}, t)$:

$$\mathbf{u}(\mathbf{x}, t) = \boldsymbol{\phi}(\mathbf{x}, \bar{\mathbf{x}}_i)\mathbf{a}(t) \tag{8.36}$$

where $\boldsymbol{\phi}(.)$ is a vector of various distribution functions, and $\mathbf{a}$ is a vector of corresponding amplitudes.

$$\boldsymbol{\phi}(\mathbf{x}, \bar{\mathbf{x}}_i) = \{ w_1(\mathbf{x}, \bar{\mathbf{x}}_1)\boldsymbol{\phi}_l(\mathbf{x}, \bar{\mathbf{x}}_1) \cdots w_Q(\mathbf{x}, \bar{\mathbf{x}}_Q)\boldsymbol{\phi}_l(\mathbf{x}, \bar{\mathbf{x}}_Q) \} \tag{8.37}$$

$$\mathbf{a}(t) = \{ \mathbf{a}_1(t) \cdots \mathbf{a}_Q(t) \} \tag{8.38}$$

Now, making use of Eq. (8.37), the optimization problem to solve for a total of NQ amplitude variables associated with various distribution functions can be defined as follows:

$$\min : \| \boldsymbol{\Phi}(\mathbf{x}, \bar{\mathbf{x}}_i)\mathbf{a} \|_p \tag{8.39}$$

subject to

$$\mathbf{H}\mathbf{a} = \mathbf{v} \tag{8.40}$$

$$\mathbf{u}_l \leq \boldsymbol{\Phi}(\mathbf{x}, \bar{\mathbf{x}}_i)\mathbf{a} \leq \mathbf{u}_u \tag{8.41}$$

where $\mathbf{H} = \mathbf{B}\boldsymbol{\Phi}(.)$ and $\boldsymbol{\Phi}(.) \in \mathcal{R}^{p \times NQ}$ is given by the following equation:

$$\boldsymbol{\Phi} = \begin{bmatrix} \phi_1(\mathbf{x}_1, \bar{\mathbf{x}}) & \phi_2(\mathbf{x}_1, \bar{\mathbf{x}}) & \cdots & \phi_{NQ}(\mathbf{x}_1, \bar{\mathbf{x}}) \\ \phi_1(\mathbf{x}_2, \bar{\mathbf{x}}) & \phi_2(\mathbf{x}_2, \bar{\mathbf{x}}) & \cdots & \phi_{NQ}(\mathbf{x}_2, \bar{\mathbf{x}}) \\ \vdots & \vdots & \ddots & \vdots \\ \phi_1(\mathbf{x}_p, \bar{\mathbf{x}}) & \phi_2(\mathbf{x}_p, \bar{\mathbf{x}}) & \cdots & \phi_{NQ}(\mathbf{x}_p, \bar{\mathbf{x}}) \end{bmatrix} \tag{8.42}$$

Depending upon the norm selected, the above algebraic optimization problem can be easily solved for the finite dimensioned amplitude vector, $\mathbf{a}$, and thereby affect the control distribution. So, the problem of finding p control variables has been reduced to finding the amplitudes of NQ distribution functions. Note that $Q = 1$ corresponds to the problem of finding a global distribution surface, while as Q increases the distribution functions becomes more capable of approximating local features of the underlying distribution surface, $\mathbf{u}(\mathbf{x}, t)$. Ideally, one would like to choose N and Q such that a substantial dimensionality reduction results ($NQ \ll p$). Initially, one can start with a global distribution surface described by N distribution functions, and if a feasible solution is not found to the optimization problem described by Eqs. (8.39)–(8.41) then more local approximations can be introduced by increasing N until a feasible solution is found. It should be mentioned that the GLO-MAP process provides a zeroth level hierarchy in the control distribution and allows us to make distribution decisions at various scales. A important advantage of the GLO-MAP approach is that all of the parameters (the $\mathbf{a}_i$s in Eq. (8.35)) are contained linearly. The outline of the control distribution algorithm using the GLO-MAP process is shown in Fig. 8.3.

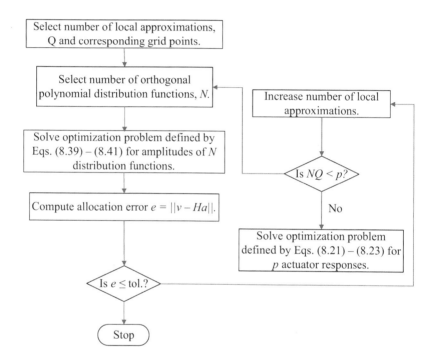

FIGURE 8.3

Flowchart for the Control Distribution Algorithm Using the GLO-MAP Orthogonal Polynomials as Distribution Functions.

Finally, it should be noted that the success of both algorithms (the RBF-based approach in Fig. 8.2 and the GLO-MAP-based approach in Fig. 8.3) depends upon how well the basis functions span the feasible solution set, $\Omega(t)$, and the performance of both algorithms is dictated by the total number of distribution functions required to have a reasonable approximation of $\Omega(t)$. Although the iterative nature of both of the algorithms seeks a good approximation of the feasible solution set, $\Omega(t)$, with a minimal number of distribution functions, there may remain cases when both algorithms fail to provide a feasible solution. We mention that the occasional failure of the distribution function approach can be attributed to the irregularity of the feasible solution set which is sometimes difficult to anticipate in high-dimensioned nonlinear problems. In the next section, we describe a hierarchical control distribution approach that makes use of these algorithms to deal with this occasional failure problem. The hierarchical approach not only allows a multi-resolution approximation of the feasible solution set, $\Omega(t)$, but also provides a mechanism for parallelizing the control distribution algorithms.

8.4 Hierarchical Control Distribution Algorithm

In the previous section we introduced the concept of distribution functions to approximate the feasible solution set, $\Omega(t)$, and showed how the use of distribution functions improves the performance of the control distribution algorithm, while keeping in check the "curse of dimensionality." However, the improvement in computational speed is generally accompanied by the degradation of the optimal solution due to errors in approximating the feasible solution set. In many cases, this degradation may not matter, especially until we get a feasible, sub-optimal solution. But, in some cases these approximation errors can lead to a situation when distribution functions fail to approximate the feasible solution set. This kind of failure may arise due to the irregularity of the feasible solution set, $\Omega(t)$. As discussed in Chapters 3 and 5, decreasing the domain of validity of different distribution functions may lead to an improvement in the approximation error, and consequently increase the region of validity of the distribution function approach. In this section, we discuss a hierarchical control distribution algorithm to improve the performance of the distribution function approximation of the feasible solution set, $\Omega(t)$, with the goal of computing the feasible solution, if it exists, with minimal computation. We mention that in the worst case scenario, the proposed algorithm requires as many resources as any other conventional control distribution algorithm that does not enjoy the possibility of model reduction.

The proposed hierarchical method decomposes the large-scale control distribution problem into many regional control distribution problems to compromise the need for real-time computation with departure from optimality. We mention that the proposed algorithm is inspired by the work of Fromherz et al. [123], Luntz et al. [115] and Jackson et al. [124]. The main steps of the algorithms can be summarized as follows:

1. Group spatially distributed actuators to generate a finite number (say, G) of small scale (regional) subsets to take advantage of regionally correlated control input distributions.

2. Combine the effects of the actuators in a particular group to form an aggregated actuator which represents all the actuators of that group. The discrete spatial coordinates associated with the aggregated actuator can be taken as the mean position (centroid) of the various actuators it represents.

3. Distribute total control effort among G aggregated actuators. In other words, assign responsibility to each aggregated actuator to produce total control effort.

4. Solve the control distribution problem recursively with the help of adaptive distribution functions in each subset.

In the previous section, we have already developed the procedure for step 4. However, the main issue with this approach is the implementation of steps 1 and 2. Basically, the problem is how to aggregate different actuators and then combine them to form an aggregated actuator for the purpose of control distribution computations. As in Refs. [115, 123, 124], a simple hierarchy scheme will involve the grouping of actuators on the basis of their spatial coordinates. If control effectiveness for each actuator is the same or is a function of spatial coordinates, then according to this approach, the actuators are grouped on the basis of their control effectiveness. For example, if we consider the example of SJAs distributed spatially over an aircraft wing, then their control effectiveness is a function of their jet frequency, strength, and the spatial position of the actuators [113]. Further, this hierarchical approach facilitates the use of distribution functions inside each subset to solve the control distribution problem recursively.

Once the actuators are grouped, then the next step is to come up with an aggregated effective actuator which conveys some averaged information about the whole group. To form an aggregated actuator by combining various actuators in a group, we need the following information about the collective response of the whole group:

1. First, the aggregated actuator needs to represent the response of all of the actuators in a particular group in some average sense.

2. Second, the constraints on the aggregated actuator should contain the information about the constraints on each individual actuator.

3. Finally, the weighting factors for each aggregated actuator's contribution to the total physical control effort is required. These weights provide an opportunity to account heuristically for the capability of each group during control distribution among each group (e.g., the type and number of actuators in each group).

The first two sets of information are easy to obtain. Usually, the response of the whole group is assumed to be the sum of the response of each individual actuator. If we let $\mathbf{u}^i = \{u_1, u_2, \cdots, u_l\}$ be a vector of the control responses of all the actuators in the ith group, then the combined/aggregated contribution of the ith group can be written as

$$\mathbf{v}^i = \mathbf{B}^i \mathbf{u}^i \qquad (8.43)$$

where $\mathbf{B}^i$ is an aggregated control distribution ("influence") matrix for the ith actuator group and consists of rows of the original $\mathbf{B}$ matrix.

$$\mathbf{B}^i = \begin{bmatrix} \mathbf{B}_1^T & \mathbf{B}_2^T & \cdots & \mathbf{B}_l^T \end{bmatrix} \qquad (8.44)$$

where $\mathbf{B}_i$ is the ith row of the control distribution matrix, $\mathbf{B}$. Further, the constraints on the aggregated group response can be found by taking the average value of the constraints on each individual actuator. Let $\bar{\mathbf{u}}_i$ and $\underline{\mathbf{u}}_i$ be vectors of upper and lower saturation limit on actuators in the ith group. Now, the constraints on aggregated response can be given as follows:

$$\mathbf{v}_l^i \leq \mathbf{v}^i \leq \mathbf{v}_u^i \qquad (8.45)$$

where

$$\mathbf{v}_l^i = \min\left(\mathbf{B}^i \underline{\mathbf{u}}^i, \mathbf{B}^i \bar{\mathbf{u}}^i\right) \qquad (8.46)$$

$$\mathbf{v}_u^i = \max\left(\mathbf{B}^i \underline{\mathbf{u}}^i, \mathbf{B}^i \bar{\mathbf{u}}^i\right) \qquad (8.47)$$

Finally, the most important part is to find the appropriate weighting function to give proper weight to the aggregated information of each group so that the required physical control effort can be divided among each group. Generally, the weighting function is chosen as the inverse of number of actuators in that particular group or depending upon some measure of controllability of the aggregated group. We choose the 2-norm of control distribution matrix as a measure of controllability for the aggregated group.

Further, let $\mathbf{v}_G$ be a vector of aggregated responses of each group, i.e., $\mathbf{v}_G = \{\mathbf{v}^1, \mathbf{v}^2, \cdots, \mathbf{v}^G\}$. Now, the aggregated response vector, $\mathbf{v}_G$, can be solved by posing the following optimization problem:

$$\min : \|\mathbf{W}\mathbf{v}_G\|_p \qquad (8.48)$$

subject to

$$\underline{\mathbf{v}}_G \leq \mathbf{v}_G \leq \bar{\mathbf{v}}_G \qquad (8.49)$$

$$\mathbf{v} = \sum_{i=1}^{G} \mathbf{v}^i \qquad (8.50)$$

where

$$\mathbf{W}_{ij} = \|\mathbf{B}_i\|\delta_{ij} \tag{8.51}$$

$$\underline{\mathbf{v}}_G = \left\{ \mathbf{v}_l^1 \ \mathbf{v}_l^2 \ \cdots \ \mathbf{v}_l^G \right\} \tag{8.52}$$

$$\bar{\mathbf{v}}_G = \left\{ \mathbf{v}_u^1 \ \mathbf{v}_u^2 \ \cdots \ \mathbf{v}_u^G \right\} \tag{8.53}$$

It should be noted that a large weighting factor $\mathbf{W}_{ii}$ causes the ith group to have a greater role in meeting the total virtual force vector $\mathbf{v}$. Further, depending upon the definition of norm in Eq. (8.48), various numerical methods as discussed in the last section can be used to find the aggregated group contribution vector $\mathbf{v}_G$.

Once the total control effort is distributed among various groups then the following optimization problem is solved using the distribution function method as discussed in the previous section.

$$\min : \|\mathbf{u}^i\|_p \tag{8.54}$$

subject to

$$\underline{\mathbf{u}}^i \leq \mathbf{u}^i \leq \bar{\mathbf{u}}^i \tag{8.55}$$

$$\mathbf{B}^i \mathbf{u}^i = \mathbf{v}^i \tag{8.56}$$

Now, we present a generic algorithm for hierarchical control allocation. The first major step of this algorithm is the task of grouping of different actuators. The grouping of actuators may be predetermined or done in real time while solving the control distribution problem. We use a hierarchical approach for the grouping of different actuators depending upon their spatial coordinates. According to the hierarchical grouping algorithm, there is only one group consisting of all of the actuators, and the control distribution problem is solved by using the RBF or the global local orthogonal polynomial distribution functions as discussed in Section 8.3. If a feasible solution is not found for the control distribution problem, then the actuators are divided into two groups. First, the required control effort is distributed among these two groups by solving the optimization problem of Eqs. (8.48)–(8.50). If a feasible solution to this optimization problem does not exist, then each group is further divided into two sub-groups, else within each group another optimization problem of Eqs. (8.54)–(8.56) is solved to compute the response of each actuator. Further, if we fail to compute a feasible solution in a particular group, then that particular group is again divided into two sub-groups. Also, if in a particular group the number of distribution functions, N, exceeds the number of actuators contained in that group, then the control distribution problem for that particular group is solved for control variables instead of amplitudes of distribution functions. This whole process is repeated recursively until a feasible solution is found or all control variables are solved for simultaneously.

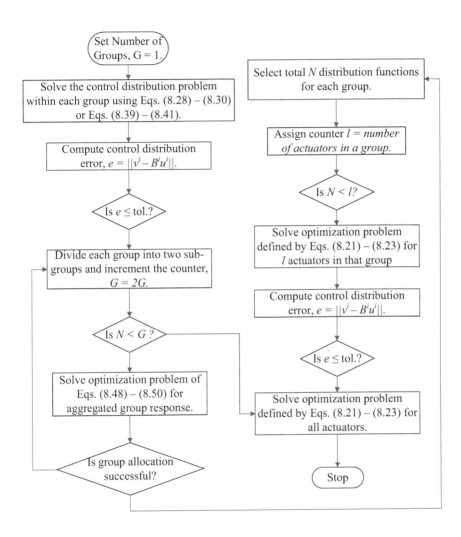

FIGURE 8.4

Flowchart for the Hierarchical Control Distribution Algorithm.

The flowchart for this hierarchical control distribution algorithm is shown in Fig. 8.4.

We mention that for real-time implementation, one might need to compromise between computational time and allocation errors. Also, based upon some previous experience or knowledge of the system, the number of groups can be predetermined and "frozen" to save some computational time. The main advantage of this hierarchical approach is that the control distribution in each group is decoupled from the distribution problem in other groups, and thus this algorithm can be highly parallelized to reduce the elapsed computation time required. In addition to this, there are many other advantages to this kind of hierarchical approach. First, the distributed nature of the actuators can be fully exploited without having the dimensionality of the optimization problem approach infinity. Second, in the case of actuator failures in one particular group, the redistribution can be adaptively performed without affecting the distribution in all groups. Finally, actuator utilization can be optimized independently for specific applications and, in principle, changed adaptively, on the fly. For example, leading edge actuation may be preferable at high Angle of Attack (AOA), while trailing edge actuation may be best for low AOA. With sufficient intelligence or rule-based logic, adaptive algorithms may be able to automatically shift emphasis of the control allocation in real time.

8.5   Numerical Results

The proposed control distribution algorithm of Fig. 8.4 is tested on a simulated control allocation problem for a virtual morphing wing embedded with millions of hypothetical actuators. Some results from these studies are presented in this section.

8.5.1   Control Allocation for a Morphing Wing

There is a significant research thrust in academics and the aerospace industry to develop advanced technologies that would enable adaptive, intelligent, shape-controllable micro and macro structures for advanced aircraft and space systems. These designs involve precise control of the shape of the structures with micro and macro level manipulations (actuation). In pursuit of these objectives, we designed a novel morphing wing prototype that can achieve an infinity of different configurations upon command. Broadly speaking, the morphing wing represents an alternative technology that adaptively shapes the flow and pressure fields over the wing by changing the curvature of the wing. This morphing technology could lead to replacement of hinged con-

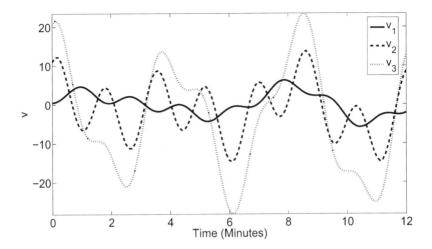

FIGURE 8.5
Physical/Virtual Control Effort.

trol surfaces, thereby achieving hingeless control. This morphing of the wing can be achieved by embedding actuators at microscales of an aerodynamic structure. The desired force and moment profile are achieved by generating moments using these actuators to deform the geometry, thereby creating a desired flow and pressure distribution over the surface.

While micro-sensors are being developed, actuation is still accomplished by a few discrete actuators. For the purpose of numerical studies, we test the idea of morphing by considering a virtual wing embedded with a 150×150 array of micro-torsional actuators to impose moments along three Cartesian coordinates. We mention that this is a hypothetical situation, but provides us a good simulation platform to test the control distribution algorithm developed in this chapter.

For simulation purposes, the control effectiveness of each actuator is assumed to be

$$\mathbf{B}_i = \begin{bmatrix} -3.7239 & 19.8465 & 15.6663 \end{bmatrix} 10^{-3} \qquad (8.57)$$

Further, each actuator is constrained to produce at max moment of $0.1\ Nm$ in either direction. To allocate the required control moments (Fig. 8.5) among the 22,500 embedded actuators, the hierarchical algorithm (shown in Fig. 8.4) is used. First, the whole control effort is assigned to all of the actuators and the response of each actuator in that regional group is approximated by orthogonal polynomials of the GLO-MAP algorithm, as discussed in Section 8.3. Figs. 8.6(a) and 8.6(b) show the plots of group allocation error and net control distribution error, respectively. From these plots it is clear that with just one

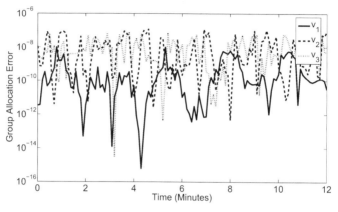

(a) Control Allocation Error Among Different Groups

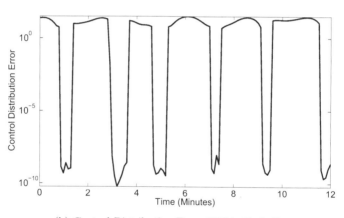

(b) Control Distribution Error Within Each Group

FIGURE 8.6
Control Distribution Results by Dividing the Actuators in 1 Group.

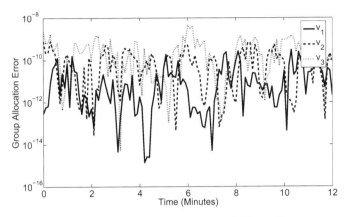

(a) Control Allocation Error Among Different Groups

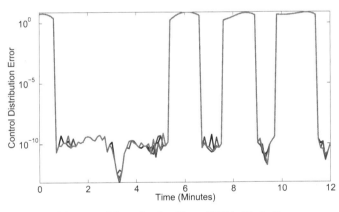

(b) Control Distribution Error Within Each Group

FIGURE 8.7
Control Distribution Results by Dividing the Actuators in 4 Groups.

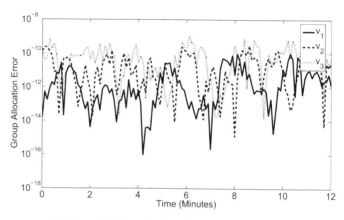

(a) Control Allocation Error Among Different Groups

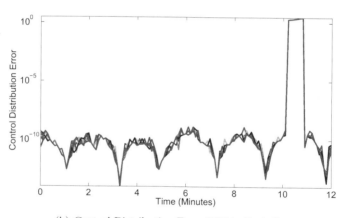

(b) Control Distribution Error Within Each Group

FIGURE 8.8

Control Distribution Results by Dividing the Actuators in 8 Groups.

group of actuators, the control distribution function approach fails to compute the feasible solution.

According to the hierarchical control distribution algorithm, we divide all actuators in 4 groups. Fig. 8.7(a) shows the plot of allocation error among the 4 groups. From this plot it is clear that the optimization problem of Eqs. (8.48)−(8.50) is solved successfully. Now, within each group, we use the orthogonal polynomial functions to distribute (interpolate) the control error among the actuators contained in that particular group. Fig. 8.7(b) shows the plot of control distribution error within each group. From this figure, it is clear that although the results have improved from the previous step, there are some instances when the distribution function approach fails to provide a solution. Similarly, Figs. 8.8(a) and 8.8(b) show the plot of control distribution among 8 different groups of actuators, and control distribution error within each group. Once again, there is an improvement from the previous step, but still there are some instances when the control distribution algorithm fails to provide a feasible solution. We repeat this process of dividing the actuators in groups recursively and finally settle down to a total of 25 groups of actuators. Fig. 8.9(a) shows the plot of control distribution error among 25 groups of actuators, whereas Fig. 8.9(b) shows the plot of control distribution error within each group. From these plots, we can conclude that the hierarchical approach performed very well in allocating the total control effort among all 22,500 actuators. We mention that 25 groups of actuators are used only in those instances when we are not able to find the solution using a lesser number of groups. We also mention that the failure of the control distribution algorithm within each group at some instant with less actuator groups can be attributed to the irregular nature of the feasible solution set. To make this point clear, we solved the control distribution problem without using distribution functions in all 25 groups. Figs. 8.10 and 8.11 show the plot of the control distribution surface for groups 1 and 25 at a particular time instant with and without making use of distribution functions, respectively. Note that the surfaces in Fig. 8.11 are highly irregular, whereas the surfaces in Fig. 8.10 are very smooth. From these plots, it is clear that the solution obtained by using distribution functions is not necessarily unique and is not guaranteed to be the optimal. However, the optimal smooth solution uses those particular distribution functions. Further, we mention that we used only six distribution functions in all of the groups and at all of the levels. That means within each group we are solving for only six amplitudes of these distribution functions. However, we need to solve for $22,500/25 = 900$ control variables if we do not use distribution functions to approximate the feasible solutions set. Fig. 8.12 shows the plot of processor time required to solve the control distribution problem with and without using distribution functions. In both methods, we use 25 groups of actuators to divide the problem into many small-scale problems. As expected, the processor time decreases significantly (two orders of magnitude) if distribution functions are used to approximate the feasible solution set. Also, a further decrease in processor time is possible

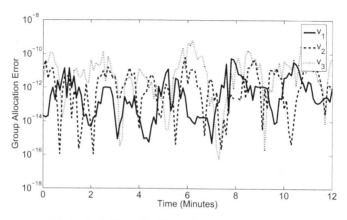

(a) Control Allocation Error Among Different Groups

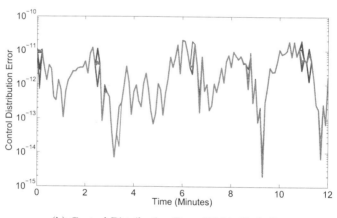

(b) Control Distribution Error Within Each Group

FIGURE 8.9

Control Distribution Results by Dividing the Actuators in 25 Groups.

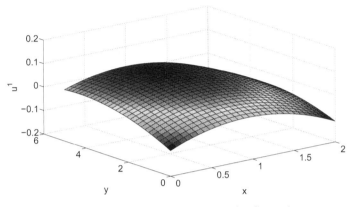

(a) Control Distribution Surface for Group 1

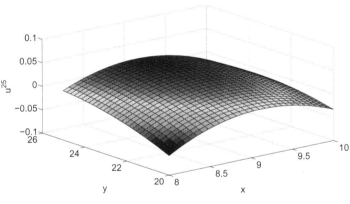

(b) Control Distribution Surface for Group 25

FIGURE 8.10
Control Distribution Surface by Using the Distribution Function Approach.

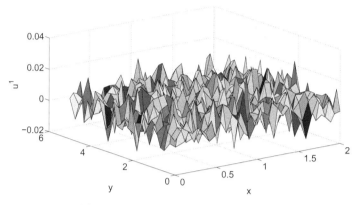

(a) Control Distribution Surface for Group 1

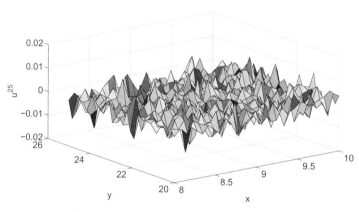

(b) Control Distribution Surface for Group 25

FIGURE 8.11

Control Distribution Surface without Using the Distribution Function Approach.

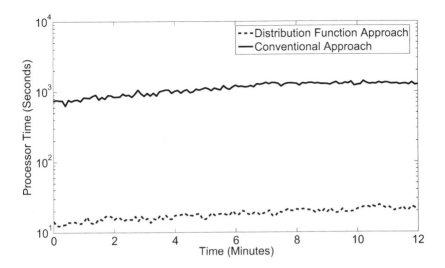

FIGURE 8.12
Processor Time to Solve the Control Distribution Problem.

by parallelized implementation of the hierarchical approach, and we mention that it is easy to parallelize this formulation. It is important that this example is a basis for optimism, but obviously one numerical study does not prove a general trend.

Finally, we mention that all numerical simulations were performed using the MATLAB [74] environment on a 1.5GHz Sony Vaio Notebook equipped with 768MB of RAM and the Microsoft Windows XP operating system. Also, we would like to mention that all optimization problems are solved using the SeDuMi [125, 126] optimization package interfaced with YALMIP [127]. SeDuMi stands for Self Dual Minimization and has been proved to solve large-scale optimization problems in an efficient manner. YALMIP is a MATLAB toolbox for rapid prototyping of optimization problems.

8.6   Summary

A general hierarchical methodology for control distribution in highly redundant actuation systems is presented in this chapter. The new method makes use of distribution functions to approximate the feasible solution set and to keep in check the "curse of dimensionality." Due to the irregular nature of the feasible solution set, it may be difficult to approximate the feasible solution set with a chosen set of smooth distribution functions. However, the approximation errors can be improved by using compactly supported distribution functions. To improve the performance of the distribution function approach, a hierarchical approach is proposed which guarantees the computation of the feasible solution if it exists. The proposed hierarchical method decomposes a large-scale control distribution problem into many small-scale control distribution problems to compromise the need for real-time computation against optimality. The main advantage of the proposed hierarchical approach is the decoupling of many small-scale problems from each other. As a consequence, the algorithm can be highly parallelized to reduce the computational burden involved. The convergence and accuracy of the proposed method are demonstrated by numerical studies. The broad generality of the method, together with simulation results, provides a strong basis for optimism for the importance and utility of these ideas. However, more testing is required to reach stronger conclusions about the utility of this algorithm.

Appendix

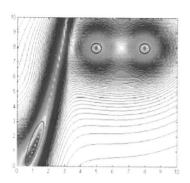 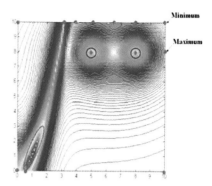

(a) Step 1: Locate Interior Extremum Points.

(b) Step 2: Locate Extremum Points along the Boundary of Input-Space.

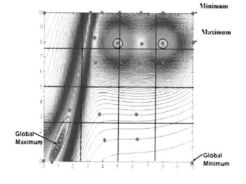 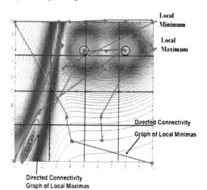

(c) Step 3: Locate Local Extremum Points.

(d) Step 4: Make a Directed Connectivity Graph of Local Extremum Points.

COLOR FIGURE 3.5

Illustration of Center Selection in the DCG Network.

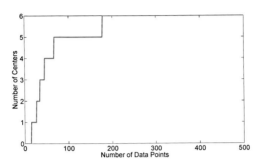

(a) Number of Centers vs. Data Points

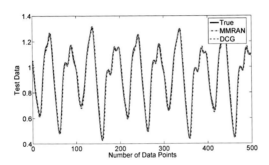

(b) Test Data Approximation Result

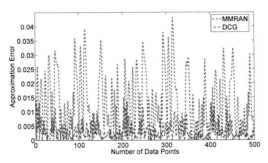

(c) Test Data Approximation Error

COLOR FIGURE 3.9

Simulation Results for Test Example 4.

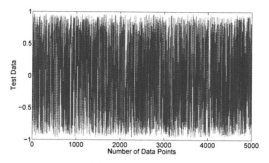

(a) True Test Data

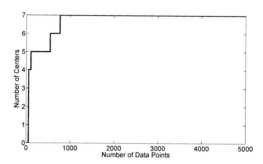

(b) Number of Centers vs. Data Points

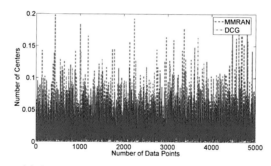

(c) Absolute Test Data Set Approximation Error

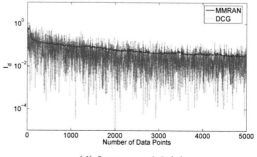

(d) Incremental $I_d(n)$

COLOR FIGURE 3.10
Simulation Results for Test Example 5.

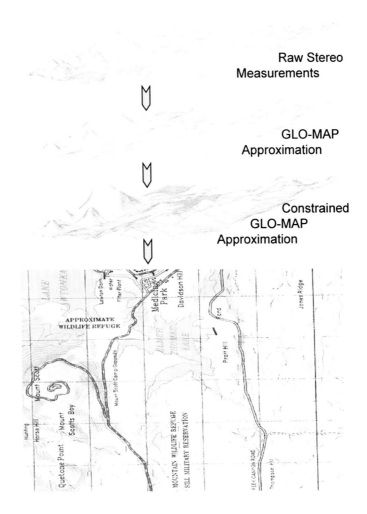

Raw Stereo
Measurements

GLO-MAP
Approximation

Constrained
GLO-MAP
Approximation

COLOR FIGURE 5.1
Approximation of Irregular Functions in 2 Dimensions.

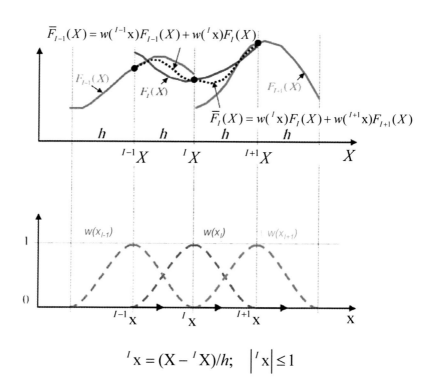

COLOR FIGURE 5.3
Weighting Function Approximation of a 1-Dimensional Function.

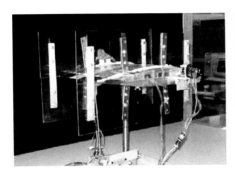

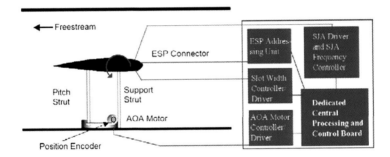

COLOR FIGURE 5.14
Hingeless Control-Dedicated Experimental Setup for Synthetic Jet Actuation Wing.

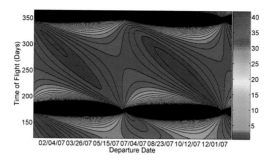

(a) True Porkchop Plot for Departure ΔV_∞

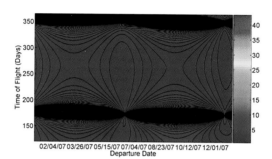

(b) True Porkchop Plot for Arrival ΔV_∞

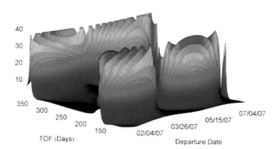

(c) True Surface Plot for Departure ΔV_∞

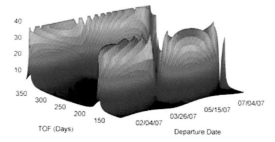

(d) True Surface Plot for Arrival ΔV_∞

COLOR FIGURE 5.21

True Departure and Arrival ΔV_∞ Plots for a Mission to the Asteroid 2003-$YN107$.

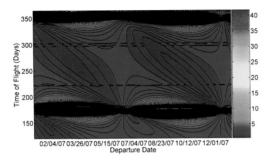

(a) Approximated Porkchop Plot for Departure ΔV_∞

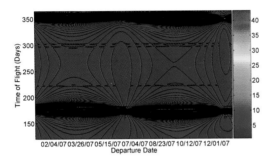

(b) Approximated Porkchop Plot for Arrival ΔV_∞

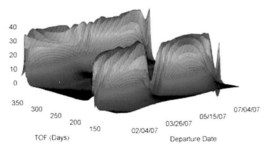

(c) Approximated Surface Plot for Departure ΔV_∞

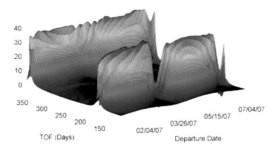

(d) Approximated Surface Plot for Arrival ΔV_∞

COLOR FIGURE 5.22

Approximated Departure and Arrival ΔV_∞ Plots for a Mission to the Asteroid 2003-YN107.

References

[1] J. L. Junkins, *Optimal Estimation of Dynamical Systems*, Sijthoff and Noordhoff International Publishers, Alphan aan den Rijn, The Netherlands, 1978.

[2] J. L. Crassidis and J. L. Junkins, *Optimal Estimation of Dynamical Systems*, vol. 2 of *Applied Mathematics and Nonlinear Science*, CRC Press Inc., Boca Raton, FL, 2004.

[3] J. L. Junkins and Y. Kim, *Introduction to Dynamics and Control of Flexible Structures*, AIAA Education Series, Reston, VA, 1993.

[4] C. S. Fraser, "Digital camera self-calibration," *Journal of Photogrammetry and Remote Sensing*, vol. 52, no. 4, pp. 149–159, 1997.

[5] K. R. Lenz and Y. R. Tsai, "Techniques for calibration of the scale factor and image center for high accuracy 3-D machine vision metrology," *IEEE Transactions on Pattern Analysis and Machine Intelligence*, vol. 10, no. 5, pp. 713–720, Sep. 1988.

[6] P. Singla, "A new attitude determination approach using split field of view star camera," M.S. thesis, Aerospace Engineering Department, Texas A&M University, College Station, TX, Aug. 2002.

[7] K. Weierstrass, "Über die analytische darstellbarkeit sogenannter willkürlicher functionen einer reellen veränderlichen," *Sitzungsberichte der Königlich Preußischen Akademie der Wissenschaften zu Berlin*, pp. 633–639, 1885.

[8] M. H. Stone, "Applications of the theory of Boolean rings to general topology," *Transactions of the American Mathematical Society*, vol. 41, no. 3, pp. 375–481, 1937.

[9] M. H. Stone, "The generalized Weierstrass approximation theorem," *Mathematics Magazine*, vol. 21, no. 4, pp. 167–184, 1948.

[10] M. H. Stone, "The generalized Weierstrass approximation theorem," *Mathematics Magazine*, vol. 21, no. 5, pp. 237–254, 1948.

[11] W. Cheney, *Analysis for Applied Mathematics*, Graduate Text in Mathematics. Springer, New York, NY, 2001.

[12] A.F. Nikiforov, S.K. Suslov, and V.B. Uvarov, *Classical Orthogonal Polynomials of a Discrete Variable*, Springer, New York, NY, 1991.

[13] A. N. Kolmogorov, "On the representation of continous function of several variables by superposition of continuous functions of one variable and addition," *Doklady Akademiia Nauk SSSR*, vol. 114, no. 5, pp. 953–956, 1957.

[14] R. O. Duda, P. E. Hart, and D. G. Stork, *Pattern Classification*, John Wiley & Sons, Inc., New York, NY, 2001.

[15] S. Haykin, *Neural Networks: A Comprehensive Foundation*, Prentice Hall, Upper Saddle River, NJ, 1998.

[16] K. Tao, "A closer look at the radial basis function networks," in *Conference record of 27th Asilomar Conference on Signals, Systems and Computers*, pp. 401–405, Pacific Grove, CA, May 1993.

[17] N. Sundararajan, P. Saratchandran, L. Y. Wei, Y. W. Lu, and Y. Lu, *Radial Basis Function Neural Networks with Sequential Learning: MRAN and Its Applications*, vol. 11 of Progress in Neural Processing, World Scientific Pub. Co., River Edge, NJ, Dec. 1999.

[18] M. Musavi, W. Ahmed, K. Chan, K. Faris, and D. Hummels, "On training of radial basis function classifiers," *Neural Networks*, vol. 5, no. 4, pp. 595–603, Jul. 1992.

[19] M. Powell, "Algorithms for approximation," in *Radial Basis Function for Multivariate Interpolation: A Review*, J. C. Mason and M. G. Cox, Eds., pp. 143–168, Clarendon Press, Oxford, 1987.

[20] J. J. Benedetto, *Harmonic Analysis and Applications*, CRC Press Inc., Boca Raton, FL, 1997.

[21] F. J. Narcowich, "Recent developments in error estimates for scattered-data interpolation via radial basis functions," *Numerical Algorithms*, vol. 39, no. 1–3, pp. 307–315, Jul. 2005.

[22] F. J. Narcowich, J. D. Ward, and H. Wendland, "Sololev bounds on functions with scattered zeros, with applications to radial basis function surface fitting," *Mathematics of Computation*, vol. 74, no. 250, pp. 743–763, Jul. 2005.

[23] F. J. Narcowich, J. D. Ward, and H. Wendland, "Refined error estimates for radial basis function interpolation," *Constructive Approximation*, vol. 19, no. 4, pp. 541–564, Aug. 2003.

[24] J. Park and I. W. Sandberg, "Universal approximation using radial basis function networks," *Neural Computation*, vol. 3, no. 2, pp. 246–257, Mar. 1991.

[25] S. Chen, C. F. N. Cowman, and P. Grant, "Orthogonal least squares learning algorithm for radial basis function networks," *IEEE Transaction on Neural Networks*, vol. 2, no. 2, pp. 302–309, Mar. 1991.

[26] V. Kadirkamanathan and M. Niranjan, "A function estimation approach to sequential learning with neural networks," *Neural Computation*, vol. 5, no. 6, pp. 954–975, Nov. 1993.

[27] J. Moody and C.J. Darken, "Fast learning in network of locally-tuned processing units," *Neural Computation*, vol. 1, no. 2, pp. 281–294, Mar. 1989.

[28] T. F. Junge and H. Unbehauen, "On-line identification of nonlinear time variant systems using structurally adaptive radial basis function networks," in *Proceedings of the American Control Conference*, Albuquerque, NM, pp. 1037–1041, June 1997.

[29] J. Platt, "A resource allocating network for function interpolation," *Neural Computation*, vol. 3, no. 2, pp. 213–225, Nov. 1991.

[30] T. Poggio and F. Girosi, "Networks for approximation and learning," *Proceedings of the IEEE*, vol. 78, no. 9, pp. 1481–1497, Sep. 1990.

[31] S. Lee and R. M. Kil, "A Gaussian potential function network with hierarchically self-organizing learning," *Neural Networks*, vol. 4, no. 2, pp. 207–224, Mar. 1991.

[32] P. Singla, K. Subbarao, and J. L. Junkins, "Direction dependent learning for radial basis function networks," *IEEE Transaction on Neural Networks*, vol. 18, no. 1, pp. 203–222, Jan. 2007.

[33] R. E. Kalman, "A new approach to linear filtering and prediction problems," *Transactions of the ASME–Journal of Basic Engineering*, vol. 82, pp. 35–45, Series D, 1960.

[34] K. Levenberg, "A method for the solution of certain nonlinear problems in least squares," *Quarterly of Applied Mathematics*, vol. 2, pp. 164–168, 1944.

[35] D. W. Marquardt, "An algorithm for least-squares estimation of nonlinear parameters," *Journal of the Society for Industrial and Applied Mathematics*, vol. 11, no. 2, pp. 431–441, 1963.

[36] N. Sundararajan, P. Saratchandran, and Y. Li, *Fully Tuned Radial Basis Function Neural Networks For Flight Control*, vol. 12 of The International Series on Asian Studies in Computer and Information Science, Kluwer Academic, Norwell, MA, 2002.

[37] IEEE Neural Networks Council Standards Committee, *Benchmark Group on Data Modeling*, http://neural.cs.nthu.edu.tw/jang/benchmark, [Online] Available, Jan. 2005.

[38] J. L. Junkins and J. R. Jancaitis, "Smooth irregular curves," *Photogrammetric Engineering*, vol. 38, no. 6, pp. 565–573, June 1972.

[39] J. O. Moody and P. J. Antsaklis, "The dependence identification neural network construction algorithm," *IEEE Transaction on Neural Networks*, vol. 7, no. 1, pp. 3–15, Jan. 1996.

[40] S. Tan, J. Hao, and J. Vandewalle, "Identification of nonlinear discrete-time multivariable dynamical system by RBF neural networks," in *Proceedings of the IEEE International Conference on Neural Networks*, vol. 5, pp. 3250–3255, Orlando, FL, 1994.

[41] M. C. Mackey and L. Glass, "Oscillation and chaos in physiological control systems," *Science*, vol. 197, no. 4300, pp. 287–289, 1977.

[42] B. A. Whitehead and T. D. Choate, "Cooperative-competitive genetic evolution of radial basis function centers and widths for time series prediction," *IEEE Transactions on Neural Networks*, vol. 7, no. 4, pp. 869–880, Jul. 1996.

[43] J. Gonzalez, I. Rojas, J. Ortega, H. Pomares, F. J. Fernandez, and A. F. Diaz, "Multiobjective evolutionary optimization of the size, shape, and position parameters of radial basis function networks for function approximation," *IEEE Transactions on Neural Networks*, vol. 14, no. 6, pp. 1478–1495, Nov. 2003.

[44] I. Rojas, H. Pomares, J. Gonzalez, E. Ros, M. Sallmeron, J. Ortega, and A. Prieto, "A new radial basis function networks structure: Application to time series prediction," in *Proceedings of the IEEE-INNS-ENNS*, vol. IV, pp. 449–454, Jul. 2000.

[45] V. Vapnik, *The Nature of Statistical Learning Theory*, Springer, New York, NY, 1995.

[46] V. Vapnik, S. E. Golowich, and A. Smola, "Support vector method for function approximation, regression estimation, and signal processing," *Advances in Neural Information Processings Systems*, vol. 9, pp. 281–287, 1997.

[47] E. W. Weisstein, "Wavelet explorer documentation: A Wolfram Web Resource," http://www.wolframresearch.com/products/applications/wavelet/contents.html, Available, Jan. 2006.

[48] E. J. Stollnitz, T. D. DeRose, and D. H. Salesin, "Wavelets for computer graphics: A primer part 2," *IEEE Computer Graphics and Applications*, vol. 15, no. 4, pp. 75–85, Jul. 1995.

[49] I. Daubechies, *Ten Lectures on Wavelets*, Number 61 in CBMS-NSF Regional Conference Series in Applied Mathematics. Society for Industrial & Applied Math, Philadelphia, PA, 1st edition, 1992.

[50] S. Mallat, *A Wavelet Tour of Signal Processing*, Number 61. Academic Press, San Diego, CA, 2nd edition, 1999.

[51] Carl De Boor, *A Practical Guide to Splines*, Springer, New York, NY, 1978.

[52] L. Piegl, *Fundamental Developments of Computer Aided Geometric Design*, Academic Press, San Diego, CA, 1993.

[53] S. N. Atluri and T. Zhu, "A new Meshless Local Petrov-Galerkin (MLPG) approach in computational mechanics," *Computational Mechancis*, vol. 22, no. 2, pp. 117–127, 1998.

[54] H. Wendland, "Local polynomial reproduction and moving least squares approximation," *IMA Journal of Numerical Analysis*, vol. 21, no. 1, pp. 285–300, 2001.

[55] D. Levin, "The approximation power of moving least squares," *Mathematics of Computations*, vol. 67, no. 224, pp. 1517–1531, 1998.

[56] T. D. Griffith, P. Singla, and J. L. Junkins, "Autonomous on-orbit calibration approaches for star tracker cameras," in *Proceedings of 2002 AAS/AIAA Spaceflight Mechanics Meeting*, San Antonio, TX, vol. 112 of Advances in Astronautical Sciences, pp. 39–51, 2002.

[57] M. A. Samaan, D. Mortari, and J.L. Junkins, "Non-Dimensional Star Identification for Un–Calibrated Star Cameras," Ponce, Puerto Rico, AAS Paper 03–131 of the AAS/AIAA Space Flight Mechanics Meeting, Feb. 9–13, 2003.

[58] B. O. Almorth, P. Stern, and F. A. Brogan, "Automatic choice of global shape functions in structural analysis," *AIAA Journal*, vol. 16, pp. 525–528, 1978.

[59] J. Goldberger, S. Roweis, G. Hinton, and R. Salakhutdinov, "Neighbourhood components analysis," *NIPS*, vol. 17, pp. 513–520, 2005.

[60] S. Mannor, I. Menache, and N. Shimkin, "Basis function adaptation in temporal difference reinforcement learning," *Annals of Operations Research*, vol. 134, pp. 215–238, 2005.

[61] R. C. Engels and J. L. Junkins, "Local representation of the geopotential by weighting orthonormal polynomials," *AIAA Journal of Guidance and Control*, vol. 3, no. 1, pp. 55–61, 1980.

[62] J. L. Junkins, P. Singla, T. D. Griffith, and T. Henderson, "Orthogonal global/local approximation in n-dimensions: Applications to input/output approximation," in *6th International Conference on Dynamics and Control of Systems and Structures in Space*, Cinque-Terre, Italy, 2004.

[63] J. L. Junkins, G. W. Miller, and J. R. Jancaitis, "A weighting function approach to modeling of geodetic surfaces," *Journal of Geophysical Research*, vol. 78, no. 11, pp. 1794–1803, 1973.

[64] Eric W. Weisstein, "Beta function," A Wolfram Web Resource, http://mathworld.wolfram.com/BetaFunction.html, Jan. 2006.

[65] P. Singla, "Multi-resolution methods for high fidelity modeling and control allocation in large-scale dynamical systems," Dissertation, Texas A&M University, College Station, TX, 126–158, May 2006,

[66] J. R. Jancaitis and J. L. Junkins, "Modeling in n dimensions using a weighting function approach," *Journal of Geophysical Research*, vol. 79, no. 23, pp. 3361–3366, 1974.

[67] J. L. Junkins and R. S. Engels, "The finite element approach in gravity modeling," *Manuscripta Geodaetica*, vol. 4, pp. 185–206, Feb. 1979.

[68] J. L. Junkins, "An investigation of finite element representations of the geopotential," *AIAA Journal*, vol. 14, no. 6, pp. 803–808, 1976.

[69] P. Singla, T. D. Griffith, K. Subbarao, and J. L. Junkins, "Autonomous focal plane calibration by an intelligent radial basis network," in *Proceedings of 2004 AAS/AIAA Spaceflight Mechanics Meeting*, vol. 119 of Advances in Astronautical Sciences, pp. 275–300, Maui, HI, 2004.

[70] P. Singla, K. Subbarao, O. Rediniotis, and J. L. Junkins, "Intelligent multi-resolution modeling: Application to synthetic jet actuation and flow control," in *42nd AIAA Aerospace Sciences Meeting and Exhibit*, number AIAA-2004-0774, Reno, NV, Jan. 5–8, 2004.

[71] W. H. Press, B. P. Flannery, S. A. Teukolsky, and W. T. Vetterling, *Numerical Recipes in C : The Art of Scientific Computing*, Cambridge University Press, Cambridge, England, 2nd edition, 1992.

[72] L. Traub, A. Miller, P. Signla, M. Tandale, J. Junkins, and O. Rediniotis, "Distributed hingeless flow control and rotary synthetic jet actuation," in *42nd AIAA Aerospace Sciences Meeting and Exhibit*, number AIAA-2004-0224, Reno, NV, Jan. 5–8, 2004.

[73] NASTRAN, http://www.mscsoftware.com.

[74] MATLAB, http://www.mathworks.com.

[75] P. Singla, T. Henderson, J. L. Junkins, and J. Hurtado, "A robust nonlinear system identification algorithm using orthogonal polynomial network," in *Proceedings of 2005 AAS/AIAA Spaceflight Mechanics Meeting*, vol. 120 of Advances in Astronautical Sciences, pp. 983–1002, Copper Mountain, CO, 2005.

[76] NEO, "Near earth object program," NASA Web Resource, http://jpl.nasa.gov, Jan. 2006.

[77] H. Schaub and J. L. Junkins, *Analytical Mechanics of Space Systems*, AIAA Education Series, Reston, VA, 2003.

[78] R. H. Battin, *An Introduction to the Mathematics and Methods of Astrodynamics*, AIAA Education Series, Reston, VA, 1999.

[79] P. A. Ionnaou and J. Sun, *Robust Adaptive Control*, Prentice Hall Inc., Upper Saddle River, NJ, 1996.

[80] K. Subbarao, M. Steinberg, and J. L. Junkins, "Structured adaptive model inversion applied to tracking aggressive aircarft maneuvers," in *AIAA Guidance, Navigation and Control Conference and Exhibit*, Montreal, Canada, August 2001.

[81] J. N. Juang and R. S. Pappa, "An eigensystem realization algorithm for modal parameter identification and model reduction," *Journal of Guidance, Control and Dynamics*, vol. 8, no. 5, pp. 620–627, 1985.

[82] A. U. Levin and K. S. Narendra, "Control of nonlinear dynamical systems using neural networks–part ii: Observability, identification, and control," *IEEE Transactions on Neural Networks*, vol. 7, no. 1, pp. 30–42, 1996.

[83] H. N. Mhaskar and N. Hahm, "Neural networks for functional approximation and system identification," *Neural Computation*, vol. 9, no. 1, pp. 143–159, 1997.

[84] K. S. Narendra and K. Parthasarathy, "Identification and control of dynamical systems using neural networks," *IEEE Transactions on Neural Networks*, vol. 1, no. 1, pp. 4–27, 1990.

[85] J. N. Juang and R. W. Longman, "Optimized system identification," Tech. Rep. TM-1999-209711, NASA, Langley Research Center, Hampton, Virginia, Oct. 1999.

[86] J. N. Juang, *Applied System Identification*, PTR Prentice Hall, Englewood Cliffs, NJ, 1994.

[87] J. N. Juang, M. Phan, L. G. Horta, and R. W. Longman, "Identification of observer/Kalman filter Markov parameters: Theory and experiments," *Journal of Guidance, Control and Dynamics*, vol. 16, no. 2, pp. 320–329, 1993.

[88] A. C. Antoulas, D. C. Sorensen, and S. Gugercin, "A survey of model reduction methods for large-scale systems," *Contemporary Mathematics*, vol. 280, pp. 193–219, 2001.

[89] A. M. Lyapunov, *The General Problem of the Stability of Motion*, Taylor & Francis, Washington, DC, English Translation of A. M. Lyapunov's work by A. T. Fuller (editor), 1992.

[90] R. A. Gingold and J. J. Monaghan, "Kernel estimates as a basis for general particle methods in hydrodynamics," *Journal of Computational Physics*, vol. 46, no. 3, pp. 429–453, 1982.

[91] T. Belytschko, Y. Y. Lu, and J. Gu, "Element free Galerkin method," *International Journal for Numerical Methods in Engineering*, vol. 37, pp. 229–256, 1994.

[92] W. Liu, S. Jun, and Y. Zhang, "Reproducing kernel particle methods," *International Journal for Numerical Methods in Fluids*, vol. 20, no. 8–9, pp. 1081–1106, 1995.

[93] I. Babuška and J. Melenk, "The partition of unity finite element method," *International Journal for Numerical Methods in Engineering*, vol. 40, no. 4, pp. 727–758, 1997.

[94] A. Duarte and J. T. Oden, "Hp clouds – an h-p meshless method," *Numerical Methods for Partial Differential Equations*, vol. 12, no. 6, pp. 673–705, 1996.

[95] T. Belytschko, Y. Krogauz, D. Organ, M. Fleming, and P. Krysl, "Meshless methods: An overview and recent developments," Tech. Rep., University of Texas, May 1996.

[96] S. N. Atluri and S. Shen, *The Meshless Local Petrov-Galerkin (MLPG) Method*, Tech Science Press, Norcross, GA, 2002.

[97] J. N. Reddy, *An Introduction to the Finite Element Method*, McGraw-Hill, Inc., New York, NY, 1993.

[98] M. Kumar, P. Singla, S. Chakravorty, and J. L. Junkins, "A multi-resolution approach to steady state uncertainty determination in non-linear dynamical systems," in *Proceedings of the 38th IEEE Southeastern Symposium on System Theory*, pp. 344–348, Cookeville, TN, March 5–7, 2006,

[99] H. Risken, *The Fokker-Planck Equation: Methods of Solution and Applications*, Springer, New York, NY, 1989.

[100] A. T. Fuller, "Analysis of nonlinear stochastic systems by means of the Fokker-Planck equation," *International Journal of Control*, vol. 9, pp. 6, 1969.

[101] D. C. Polidori and J. L. Beck, "Approximate solutions for nonlinear vibration problems," *Probabilistic Engineering Mechanics*, vol. 11, pp. 179–185, 1996.

[102] P. W. Chodas and D. K. Yeomans, "Orbit determination and estimation of impact probability for near earth objects," in *Proceedings of the Guidance and Control*, vol. 101 of *Advances in the Astronautical Sciences*, pp. 21–40, 1999.

[103] M. C. Wang and G. Uhlenbeck, "On the theory of Brownian motion ii," *Reviews of Modern Physics*, vol. 17, no. 2–3, pp. 323–342, 1945.

[104] P. Singla and T. Singh, "A Gaussian function network for uncertainty propagation through nonlinear dynamical system," *18th AAS/AIAA Spaceflight Mechanics Meeting*, Galveston, TX, Jan. 27–31, 2008.

[105] M. Kumar, P. Singla, S. Chakravorty, and J. L. Junkins, "The partition of unity method to the solution of the Fokker-Planck equation," in *AIAA/AAS Astrodynamics Specialist Conference*, Keystone, CO, Aug. 21–24, 2006.

[106] G. Muscolino, G. Ricciardi, and M. Vasta, "Stationary and non-stationary probability density function for non-linear oscillators," *International Journal of Non-Linear Mechanics*, vol. 32, pp. 1051–1064, 1997.

[107] M. D. Paola and A. Sofi, "Approximate solution of the Fokker-Planck-Kolmogorov equation," *Probabilistic Engineering Mechanics*, vol. 17, pp. 369–384, 2002.

[108] A. Kunert, "Efficient numerical solution of multidimensional Fokker-Planck equation with chaotic and nonlinear random vibration," in *Vibration Analysis − Analytical and Computational*, T. C. Huang et al., Ed., Conference on Mechanical Vibration and Noise, pp. 51–60, Miami, FL, 1991.

[109] S. F. Wojtkiewicz, L. A. Bergman, and B. F. Spencer, Jr., "High fidelity numerical solutions of the Fokker-Planck equation," in *Proc. of ICOSSAR'97: The 7th International Conference on Structural Safety and Reliability*, A. Bazzani, J. Ellison, H. Mais, and G. Turchetti, Eds., Kyoto, Japan, Nov. 24–28, 1997.

[110] R. S. Langley, "A finite element method for the statistics of random nonlinear vibration," *Journal of Sound and Vibration*, vol. 101, no. 1, pp. 41–54, 1985.

[111] E. A. Johnson, S. F. Wojtkiewicz, L. A. Bergman, and B. F. SpencerJr., "Finite element and finite difference solutions to the transient Fokker-Planck equation," in *Proc. of a Workshop: Nonlinear and Stochastic Beam Dynamics in Accelerators − A Challenge to Theoretical and Numerical Physics*, A. Bazzani, J. Ellison, H. Mais, and G. Turchetti, Eds., Lüneburg, Germany, 1997.

[112] A. Masud and L. A. Bergman, "Application of multi-scale finite element methods to the solution of the Fokker-Planck equation," *Computer Methods in Applied Mechanics and Engineering*, vol. 194, pp. 1513–1526, 2005.

[113] A. Miller, L. Traub, O. Redeniotis, P. Singla, M. Tandale, and J. L. Junkins, "Distributed hingeless flow control and rotary synthetic jet actuation," in *42nd AIAA Aerospace Sciences Meeting and Exhibit*, number AIAA-2004-224, Reno, NV, Jan. 5–8, 2004.

[114] O. Härkegård, "Backstepping and control allocation with applications to flight control," Ph.D. dissertation, Linköping University, SE-581 83, Linköping, Sweden, 2003.

[115] J. E. Luntz, W. Messner, and H. Choset, "Distributed manipulation using discrete actuator array," *The International Journal of Robotics Research*, vol. 20, no. 7, pp. 553–583, Jul. 2001.

[116] R. Fletcher, *Practical Methods of Optimization*, 2nd edition, John Wiley & Sons, New York, NY, 1987.

[117] S. Boyd and L. Vandenberghe, *Convex Optimization*, Cambridge University Press, New York, NY, Mar. 2004.

[118] G. B. Dantzig, *Linear Programming and Extensions*, Princeton University Press, Princeton, NJ, 1963.

[119] R. H. Bartels and G. A. Golub, "The simplex method of linear programming using LU decomposition," *Communications of the ACM*, vol. 12, no. 5, pp. 266–268, 1969.

[120] S. J. Wright, *Primal-Dual Interior-Point Methods*, Society for Industrial and Applied Mathematics, Philadelphia, PA, 1997.

[121] K. A. Bordignon, "Constrained control allocation for systems with redundant control effectors," Ph.D dissertation, Virginia Polytechnic Institute and State University, Blacksburg, VA, 1996.

[122] J. M. Buffington and D. F. Enns, "Lyapunov stability analysis of daisy chain control allocation," *Journal of Guidance, Control, and Dynamics*, vol. 19, no. 6, pp. 1226–1230, Nov.–Dec. 1996.

[123] M. P. Fromherz and W. B. Jackson, "Force allocation in a large scale distributed active surface," *IEEE Transaction on Control Systems Technology*, vol. 11, no. 5, pp. 641–655, Sep. 2003.

[124] W. B. Jackson, M. P. J. Fromherz, D. K. Biegelsen, J. Reich, and D. Goldberg, "Constrained optimization based control of real time largescale systems: Airjet object movement system," in *Proceedings of the 40th International Conference on Decision and Control*, vol. 5, pp. 4717–4720, Orlando, FL, Dec. 2001.

[125] J. F. Sturm, "Using SeDuMi 1.02, a MATLAB toolbox for optimization over symmetric cones," *Optimization Methods and Software*, vol. 11, no. 12, pp. 625–653, 1999.

[126] E. Andersen, J. Gondzio, C. Meszaros, and X. Xu, *Interior point methods for mathematical programming*, Chapter 6, Implementation of Interior Point Methods for Large Scale Linear Programming, pp. 189–252, Kluwer Academic Publishers, Norwell, MA, 1996.

[127] J. Löfberg, "YALMIP: A toolbox for modeling and optimization in MATLAB," in *Proceedings of the CACSD Conference*, Taipei, Taiwan, 2004.

Index